CONSTRUCTION MANAGEMENT

Emerging Trends and Technologies

Trefor P. Williams, Ph.D., P.E.
Department of Civil and Environmental Engineering
Rutgers University

Australia • Brazil • Japan • Korea • Mexico • Singapore • Spain • United Kingdom • United States

DELMAR
CENGAGE Learning™

Construction Management: Emerging Trends and Technologies
Trefor Williams

Vice President, Career and Professional Editorial: Dave Garza

Director of Learning Solutions: Sandy Clark

Acquisitions Editor: James DeVoe

Managing Editor: Larry Main

Product Manager: Mary Clyne

Editorial Assistant: Cris Savino

Vice President, Career and Professional Marketing: Jennifer McAvey

Marketing Director: Deborah Yarnell

Marketing Manager: Jimmy Stephens

Associate Marketing Manager: Mark Pierro

Production Director: Wendy Troeger

Production Manager: Mark Bernard

Art Director: Bethany Casey

Library of Congress Control Number: 2009933141

ISBN-13: 978-1-4283-0518-2

ISBN-10: 1-4283-0518-1

Delmar
5 Maxwell Drive
Clifton Park, NY 12065-2919
USA

Cengage Learning is a leading provider of customized learning solutions with office locations around the globe, including Singapore, the United Kingdom, Australia, Mexico, Brazil, and Japan. Locate your local office at:
international.cengage.com/region

Cengage Learning products are represented in Canada by Nelson Education, Ltd.

For your lifelong learning solutions, visit **delmar.cengage.com**

Visit our corporate website at **cengage.com.**

Printed in the United States of America
1 2 3 4 5 6 7 12 11 10 09 08

Contents

Preface

This text is written for students of construction management, construction technology, and civil engineering who are taking their first course in construction management. The goal of the book is to provide a foundation in the important aspects of construction management and suggest new and emerging areas that will concern the construction manager in the future.

This book addresses the unique features of construction management for the 21st century as it provides an overview of the construction industry and the management of construction projects. *Construction Management: Emerging Trends and Technologies* offers solid, foundational concepts in traditional areas, including construction contracts, cash flow, estimating, and scheduling. To set this text apart from traditional books on the subject, however, the content moves beyond traditional areas to explore emerging areas of interest, such as megaprojects, design-build construction, public–private partnerships, the application of information technology to construction, configuration management, and sustainable construction. This book can be used as a resource for anyone seeking a better understanding of the up-and-coming state of the construction management industry.

The material in the text and its structure is to a large extent influenced by my experience in construction projects and teaching at Rutgers University for 21 years.

My experience in the industry has led me to believe that the fundamental place to begin a study of construction is to understand how contracts affect all aspects of a construction project. Therefore, the book opens with extensive coverage of construction contractual relationships. I have attempted to explain contracts with clarity for those new to construction. In addition, my research has led me to believe that an introduction to the possibilities of applying computers and information technology to construction issues is essential for students entering the workforce in the 21st century. Therefore, I have attempted to infuse the text with applications of information technology. I have tried to explain all topics in clear language to help students who are new to construction management understand topics like scheduling and cash flow.

Case studies of various projects are included in the text. They are included to amplify the material in the book, and to provide a "real world" link to the topics discussed. Shorter field notes are also contained in the text. Their purpose is to provide additional information or background about a topic in the text.

Features:

- Illustrations, tables, and charts supplement key information
- Scheduling coverage includes both the basics of the CPM method and the use of computers, offering traditional as well as modern views of this key construction management concept
- Coverage of estimating offers discussions of both Web-based estimating databases and the use of computer programs, emphasizing the increasing role of technology in the field
- An introduction to building information modeling and the application of 4D models in construction
- Discussion of configuration management and its application to the construction industry
- Extensive coverage of construction contracts
- Discussion of emerging contractual forms like public–private partnerships
- A chapter on green and sustainable construction
- Description of the application of information technology to construction projects

Acknowledgments

I would like to acknowledge the encouragement of the director for the Center for Advanced Infrastructure and Transportation at Rutgers University, Ali Maher, to write this book. Most important, I would like to thank my wife Nancy for her support and encouragement during the writing of this book.

In addition, the publisher wishes to acknowledge the valuable comments provided by the following reviewers throughout the development process:

Jim Carr, University of Arkansas at Little Rock
Daphene Koch, Purdue University
William Moylan, Eastern Michigan University
John Schaufelberger, University of Washington
Kenneth Sullivan, Arizona State University

About the Author

Trefor P. Williams, Ph.D., P.E., is a professor of civil and environmental engineering at Rutgers University, and a registered professional engineer in New York and New Jersey. He is the author of *Information Technology for Construction Managers, Architects and Engineers*, © 2007, Delmar Cengage Learning.

chapter 1

The Nature of the Construction Industry

Chapter Outline

Introduction

We rely on the fabrication and renewal of facilities for all aspects of our lives. Industrialized economies require construction to renew and build new infrastructure. *Infrastructure* can be defined as the basic facilities and installations needed for the functioning of society. These facilities include everything from roads and bridges to schools and hospitals. Besides infrastructure, we rely on the construction industry to provide new dwellings as our population increases.

Examples of construction can be found everywhere. New homes are constructed, and existing homes are remodeled. When a chemical company sees new opportunities in refining ethanol fuels, it requires a new plant in which to do so. All our transportation needs are met by the construction of highways, airports, and bridges. Buildings must be constructed to house offices, schools, and hospitals. Thus, any residential, industrial, or commercial facility you can imagine was built as a construction project.

The increasing complexity and magnitude of construction projects require that builders, construction managers, and engineers understand the latest trends and techniques to tackle increasingly difficult projects in the most efficient manner. The goal of this book is to discuss the important issues facing construction contractors and to suggest effective methods of managing and controlling construction companies and projects.

Construction is a vast industry in the United States and throughout the world. Huge sums of money are expended providing new facilities or **retrofitting** existing facilities to satisfy the needs of society for shelter and economic growth. Construction projects can be very small, such as a home renovation. However, some construction projects can be enormous such as the construction of a subway system for a major city, a large suspension bridge, or a 100-story skyscraper. With the increasing complexity of our society and emergence of technical advances made in construction equipment and techniques, there has been a corresponding increase in the size of the construction projects undertaken. Very large-scale construction projects are becoming increasingly commonplace, but the management of these projects is demanding and requires knowledge of finance, estimating, and planning to be effectively controlled. Examples include the Palm Islands project in Dubai and the New York City Second Avenue Subway.

Some Statistics About the Construction Industry

Statistics about the construction industry can indicate its size and importance to the economy. Not only is construction a major component of the U.S. economy, but it is also a major employer. Here are some examples that indicate the importance of the construction industry in the United States:

- Construction accounted for approximately 4.1% of the **gross domestic product (GDP)** in 2007 (Lindberg and Monaldo, 2008). The GDP is the total value of goods and services produced by a nation in a given year. The total dollar amount of construction was of the order of $1.2 trillion in 2007 (Census Bureau, 2009a).

- Residential construction accounted for 45% of the $1.2 trillion dollar total. Nonresidential construction accounted for 30% of the total spending on construction, whereas government projects including large infrastructure projects accounted for 25%. Nonresidential private construction includes commercial buildings, stores, shopping centers, warehouses, manufacturing plants, private health care facilities, and power plants. Public construction includes the nation's transportation infrastructure of roads, airports, and bridges, as well as water delivery and treatment systems, wastewater systems, schools, and parks.
- Construction of over 1.36 million privately owned housing units was initiated in the United States in 2007 (U.S. Census Bureau, 2009b).
- In 2005, the construction industry employed approximately 6.8 million workers.

Projects and the Construction Industry

Construction is typically carried out as a **project**. What do we mean when we discuss a "*project*"? There are many definitions. Certainly, construction is not the only industry in which projects are carried out. In the software industry the development of a new computer program may be referred to as a "project." In the aerospace industry, the development of a new aircraft may be viewed as a "project." Some definitions of a project are the following:

- A temporary endeavor undertaken to create a unique product. Temporary means that the project has a definite end date. Unique means that the product created is different from other projects undertaken by the same construction organization (Project Management Institute, 2004).
- A unique venture with a beginning and an end, undertaken to meet established goals within the constraints of time and resources.

The emphasis on projects in the construction industry sharply differentiates it from manufacturing industries. In manufacturing, the emphasis is on selling as many standardized manufactured items as possible, whereas, in construction, the emphasis is on constructing individual, unique projects that may be long lived.

Comparison of Construction Projects and Automobile Manufacturing

The automobile industry is an important component of most industrialized economies. It is interesting to consider the differences between the automobile and construction industries. Table 1.1 illustrates some of them.

The construction industry is much more diverse than the automobile industry. Management techniques that may be applicable to the controlled environment of an automobile assembly plant are difficult to apply to the project-oriented construction industry.

Table 1.1 Comparison of construction and automobile manufacturing

CONSTRUCTION INDUSTRY	AUTOMOBILE MANUFACTURING INDUSTRY
Builds unique facilities with requirements defined by an owner.	Builds a limited number of models with specifications determined by a manufacturer. Replicate same model many times.
Typically, can participate only in a limited portfolio of projects at any time. Projects vary widely in scope and complexity. Some projects are long lived.	Produces and sells many automobiles in a single year.
Projects are conducted outdoors in demanding environments.	Automobiles in plants located at sites of the manufacturer's choosing. The fixed plant site lends itself to the automation of fabrication equipment.
Raw construction materials must be transported to individual construction sites from many different locations and fabricated there.	Raw materials are brought to the central location of the plant for fabrication.
Construction companies vary widely in size from single-person firms engaged in home remodeling to large construction companies that undertake complex infrastructure and industrial projects. Because of the large difference in firm size and behavior, making generalizations about the construction industry is difficult.	Barriers to entry in the industry are large due to its capital-intensive nature. Ten to fifteen global corporations dominate the industry. All the global firms demonstrate similar behavior in the way they manage their vast undertakings.

Construction Project Participants

When studying construction, it is important that we understand the various participants in a construction project. A construction project brings together a diverse team of people with different areas of expertise and outlooks on the conduct of the projects. In a basic construction arrangement the important parties in a project are the following:

- **Owner.** The owner is a person, company, or agency for which a project is being constructed. The owner pays for the project and specifies the project requirements. An owner may be an individual needing a new home, a developer building a shopping center, or a state department of transportation needing a new road for its citizens.
- **Designer.** An individual, a single design firm, or a group of different firms can design a construction project. The designer typically produces plans and specifications for what is to be constructed. For a residence, the designer may be a single architect. For a large skyscraper, the designer may be a team of different firms that includes architects, structural engineers to design the facilities structural components, mechanical engineers to design the heating and air conditioning, and traffic engineers to design site access and parking.
- **Construction contractor.** The firm that constructs the project is a construction contractor. In traditional construction contracts, the contractor is typically not

involved in the design but is contracted by the owner to construct on the basis of the design developed by the designer.

- **Subcontractors.** Construction projects can be extremely complex, and the prime contractor may not have the expertise to complete all the required project tasks. Therefore, the prime contractor often hires subcontractors to complete portions of the work. Coordination of the activities of many different subcontractors on a large project can be a difficult task for the prime construction contractor.
- **Government agencies.** Construction projects are increasingly subject to greater scrutiny because of increased environmental concerns. Governmental agencies are often the owners of large projects. Yet, it is often necessary for owners of construction projects to get permission from government agencies to proceed with construction. For large projects, getting permission from government agencies often takes years of planning studies and environmental impact statements before actual construction can begin. Naturally, governments require these studies to protect the interests of citizens to ensure that a project is beneficial to the overall community.
- **Community.** Construction projects are increasingly conducted in a manner that minimizes impacts to the surrounding community. In our increasingly urbanized world, people are sensitive to traffic delays or noise that is caused by a construction project. Therefore, construction contractors are often asked to find innovative ways of conducting construction that minimize impacts to the neighboring community. It is not uncommon to see restrictions placed on contractors when they work on highways projects or conduct underground blasting activities.

Specialization Within the Construction Industry

Construction encompasses a broad range of activities. These activities range from renovating and building homes to constructing large infrastructure projects such as power plants, dams, and highways. The industry is so vast that it is impossible for any single construction company to amass the expertise to participate in all types of construction projects. Construction companies specialize in certain segments of the construction industry and can generally be classified into the following broad categories:

- Residential construction
- Commercial building construction
- Industrial construction
- Heavy construction

Residential contractors vary widely in size, range from small companies building single homes to large national or multinational companies developing large residential complexes of homes and condominiums.

Commercial building construction includes the construction of skyscrapers, office buildings, warehouses, schools, stadiums, and hospitals. Turner Construction, one of the largest commercial construction companies in the United States, provides a link on its Web site (www.turnerconstruction.com/) to a graphical Web page called *Turner City*, which shows a drawing of all the projects (arranged as a city) it has completed in the previous year. This Web page shows the broad scope of work that a large building contractor can undertake. Smaller commercial building contractors will specialize in smaller niches of the building category, sometimes specializing in hospitals or schools.

Industrial construction involves the construction of power plants, chemical plants, or oil refineries. Heavy construction contractors build highways, dams, airports, and the infrastructure our society requires for essential services such as transportation and water supply. Heavy and industrial contractors tend to be large firms that are capable of undertaking the large-scale projects that are common in these areas.

Building Codes

Building codes are regulations concerning how buildings must be designed to protect the health, safety, and welfare of the public. New construction must conform to the latest version of the building codes. The origins of modern building codes lie in the need for improved building fire protection systems after fires such as the Chicago fire of 1871 (Ching and Winkel, 2007).

In building and industrial construction, the designer is responsible for ensuring that the design conforms to the applicable building codes. In addition to the building codes, the designer must be aware of specialized fire, electrical, and plumbing codes that apply. In residential construction, an architect or engineer may not be involved; the contractor building the home will be responsible for conformance with the building codes.

Many governments in the United States have their own building code, however, typically based on a model code called the International Building Code. Homes and town houses less then three stories in height must comply with the International Residential Code. State and local jurisdictions modify this code to their own particular requirements.

Emerging Trends in the Construction Industry

Several recent trends have emerged in the construction industry. First, as we have already mentioned, there has been a tendency to build bigger and more complicated projects, often called megaprojects. Second, the construction expertise required to construct megaprojects requires large construction contractors that can undertake projects anywhere in the world. Thus, the construction industry is becoming increasingly globalized.

Megaprojects

Megaprojects are typically huge infrastructure projects that require billions of dollars and several years to complete. They involve many different parties including members of the affected communities, government agencies, the owner, designers, construction contractors, subcontractors, and material suppliers and fabricators. Megaprojects can be highly controversial. Flyvbjerg et al. (2003) have discussed how many transportation megaprojects have historically been constructed for much higher sums than their original cost estimates. Possibly, these problems occur because of overly optimistic cost estimates by project backers in the early planning stages of a project, and overstatements of project benefits. Examples include the Channel Tunnel linking England and France that had an 80% cost overrun and Boston's artery/tunnel project that had a 196% cost overrun. There is increasing pressure from taxpayers and investors to demonstrate the need for a project during the planning stages of a project and control costs on these large-scale projects during construction.

The management of such megaprojects is a complex undertaking. Often, because they are undertaken to fulfill important societal needs, there is extreme pressure to construct these large projects as rapidly as possible, while at the same time minimizing disruptions and negative environmental impacts. Rigorous planning, estimating, and scheduling procedures are required to successfully implement megaprojects in a manner that is satisfactory to the projects stakeholders. Construction managers must exercise considerable skill to complete megaprojects on time and within budgets. One of the main purposes of this book is to provide students with an introduction to the techniques that are used to manage complex projects.

There are many megaprojects in the world. Some current examples include the following:

- The construction of the Second Avenue Subway line in New York City
- The Palm Islands project in Dubai in the United Arab Emirates
- The replacement of the Bay Bridge between Oakland and San Francisco

The construction of the Second Avenue Subway line in New York City has been discussed for many years and will fulfill a need for additional mass transit infrastructure on the east side of New York City. However, construction in a large city, especially in an older area like Manhattan, provides many construction challenges. These challenges include conducting the construction in a manner that does not disrupt traffic and has minimum effect on existing private structures. Additionally, existing infrastructure such as water, telephone, electricity, and gas must be maintained during construction. The difficulties of constructing in a complex urban environment vastly increase the complexity of projects and the planning that is necessary to construct them successfully. The needs to work within the constraints of the urban environment often entail construction techniques and methods that are much more complicated than those for greenfield projects in other locations. Greenfield projects are projects that are built on a new undisturbed site. Urban projects are typically brownfield projects built at a location where previously built structures or buildings existed. Brownfield projects are typically more complicated to construct because the remnants of existing structures must be removed.

Figure 1.1 Using a tunnel boring machine to dig Second Avenue Subway

Figures 1.1 and 1.2 show two different tunneling methods that will be employed on the Second Avenue line. In some areas the tunneling is in bedrock that requires a tunnel boring machine. Figure 1.1 shows a tunnel boring machine and the conveyer belt that will be used to bring excavated rock to the surface. In other areas the subway passes through softer soils that can be excavated by opening a cut in the ground, excavating the tunnel, and then covering over the subway ceiling. The disturbed street overhead is repaved after the tunnel is constructed. Figure 1.2 shows this cut-and-cover method of tunneling. Figure 1.2 also shows the many constraints that affect the project including the need to keep traffic lanes open for motor vehicles and to avoid disturbance to nearby buildings. The subway project is complicated by being so large that the subway tunnel passes through different soil and rock types, each requiring a different equipment and construction techniques. Figure 1.3 shows the work taking place underground using the cut-and-cover technique.

The costs of building the Second Avenue Subway line will be substantial. The first phase of the project began in 2006 and is scheduled to be completed by 2012. The initial phase is estimated to cost \$3.8 billion. Subsequent sections of the project are estimated to cost \$16 billion. Another aspect of megaprojects is that they require extensive planning, environmental, and design studies before construction can begin. The Second

Figure 1.2

Cut-and-cover tunneling in constrained space

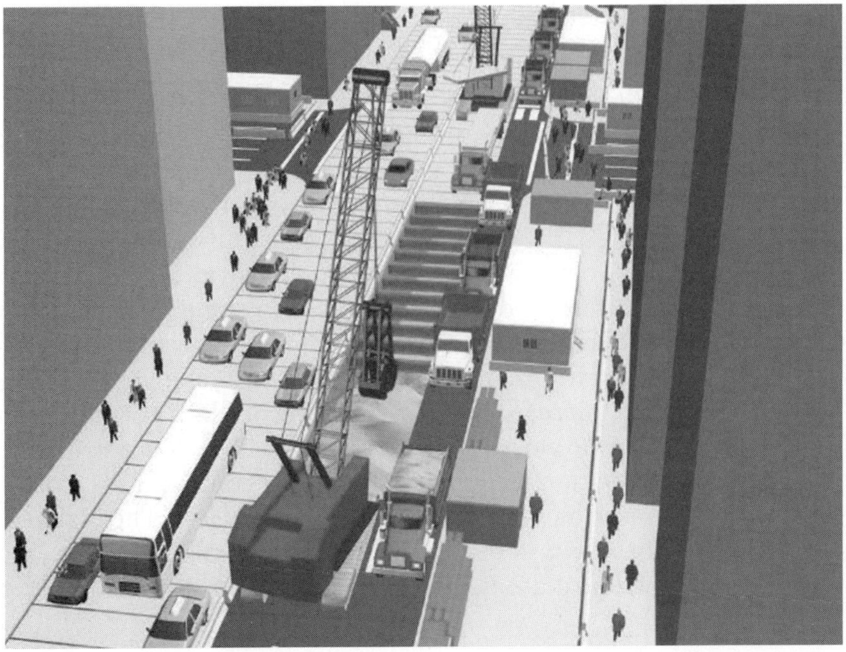

Figure 1.3 An example of work taking place underground using the cut-and-cover technique

Avenue Subway line has been discussed for 20 years, and the formal planning process was initiated in 1999 and not completed until 2005.

Megaprojects take place all over the world. American construction companies often participate in international projects. The Palm Islands development is an example of an international megaproject, which consists of three-palm shaped islands that are being constructed off the coast of Dubai in the United Arab Emirates. When completed, the project will increase Dubai's shoreline by 325 miles and will include over 100 luxury hotels, 10,000 exclusive residential beachside homes, 5,000 apartments, marinas, shopping centers, and restaurants. The project includes the entire infrastructure required to serve a community including roads, a monorail system, electric power substations, and water-treatment facilities. The project is being built by many different international construction contractors with specialized knowledge in dredging and marine work, with heavy contractors to construct the islands infrastructure and transportation facilities, and building contractors to construct the hotels, homes, and apartment buildings (Nakheel L.L.C., 2006). The Nakheel property development company has the task of coordinating this complex project. Its Web site, www.nakheel.com, provides additional information about this project.

Globalization

Construction has become a global industry. Large construction contractors now work in a global environment, constructing projects simultaneously in several different countries. This development is part of a larger trend affecting the world. This trend is called **globalization.** *Globalization* can be defined as the situation where international borders become increasingly irrelevant, and economic interdependencies between nations are heightened. Globalization can have negative impacts, by accentuating differences in social cultures and business practices (Ngowi et al., 2005).

There are many challenges involved in constructing projects where business practices and expectations are different. A construction contractor undertaking international projects must consider several risk factors (Bauml, 1997):

- **Local construction practices.** Construction equipment and practices may vary from nation to nation.
- **Local business practices.** Applicable contracts, labor agreements, regulations, and laws vary from country to country.
- **World and national economic conditions.** The economic situation of a country may make it a risky place to do business.
- **Levels of spending by the local government.** When considering entering a particular national market, a global contractor must consider the amount of money being spent on infrastructure projects by the local government.
- **Priorities for construction spending of the host nation.** A contractor with a particular area of expertise must determine that there is a market for its services in the host country.

CASE STUDY: THE SAN FRANCISCO-OAKLAND BAY BRIDGE SEISMIC SAFETY PROJECT

The San Francisco-Oakland Bay Bridge Seismic Safety project is another example of a complex megaproject. The Loma Pietra earthquake severely damaged the existing bridge. Engineers determined that the eastern portion of the 8.4-mile long bridge needed to be replaced and that the western portion needed to be retrofitted to withstand expected earthquakes.

Because of its complexity the project has been divided into several construction contracts. These contracts are so large that they are being undertaken by a consortium of several large construction contractors. The total cost of all of the contracts is estimated to be about $6.3 billion, and the completion date of the projects is estimated to be in 2014. Construction of the east bridge skyway was begun in 2002. This construction illustrates the huge costs involved and the long-term nature of this complex project. Figures 1.4–1.6 show the construction of the new east bridge skyway. The pictures illustrate the large scope of the project with many diverse activities occurring simultaneously. In particular, note the large number of construction cranes required to assemble the bridge and bridge deck, as well as the need for a large crane on a barge to drive piles.

Photo by John Huseby.

Figure 1.4 Building New Bay Bridge. *Courtesy of the Department of Transportation*

Photo by John Huseby.

Figure 1.5 View of Bridge Piers. *Courtesy of the Department of Transportation*

Photo by John Huseby.

Figure 1.6 Driving Piles for Bridge. *Courtesy of the Department of Transportation*

The project has some unique and technically advanced features, which include the construction of a self-anchored suspension bridge in the eastern portion of the project where a web of cables attached to a tower wrap around and cradle the roadway. The complexity of a project of this type is further underscored because there was considerable controversy concerning the appropriate design for the suspension bridge due to fears that it would cost too much to construct. A causeway structure was considered as an alternative. Construction of this self-anchored suspension span is the largest infrastructure contract ever awarded in California.

It is also interesting to note that some aspects of the project were shaped by the desires of the user community. The original bridge was a two-level structure. However, during the design stage of the project, citizens of the San Francisco Bay area requested that the roadways be placed side by side to provide a more open feeling and better views of the cities and bay. The new portions of the bridge will have this configuration, rendering the construction more complex because a transition section will need to be constructed between the older two-level bridge and the new bridge. Environmental concerns often add to the complexity of megaprojects. In the case of the Bay Bridge, fear of negative impacts on the fish population requires that portions of the old bridge (to be demolished) must be taken apart piece by piece. Although bridges are demolished using explosives, the concerns of government agencies and citizens require a more environmentally friendly but a more time-consuming process. This complex bridge-rehabilitation project illustrates the many factors that affect a complex project including the need to keep this important bridge open to traffic during reconstruction, the aesthetic and environmental concerns of the citizens, the need to minimize costs, and the need for large construction contractors to form teams to manage a project so large in scope.

There are some controversial issues involved in the globalization of the construction industry. Globalization has caused construction companies in the developed world to become even larger as they expand their operations globally. These contractors are typically based in the United States, western Europe, Japan, or South Korea. Controversy arises because it is often believed that the global contractors stunt the development of the construction industry in developing nations (Ngowi et al., 2005). To overcome the fears of local contractors in developing nations, global contractors often enter into partnering arrangements with contractors in the developing nation.

The U.S. Transportation Infrastructure Crisis

In any industrialized nation a smoothly functioning transportation **infrastructure** of highways, bridges, railways, waterways, and airports is required to keep the economy going. *Infrastructure* can be defined as the arrangement of a nation's transportation networks and facilities. Recently, a belief has emerged in the United States that we are not renewing and expanding our transportation infrastructure to keep pace with increasing demands for transportation. This crisis presents both an opportunity and a challenge to the construction community to build new facilities and to revitalize worn-out roads and bridges. The American Society of Civil Engineers (ASCE) has stated that congested highways, deteriorating bridges, and overflowing sewers are reminders of the crisis that jeopardizes U.S. prosperity and quality of life (ASCE, 2005). The ASCE infrastructure report card gives the U.S. infrastructure a grade of D. In 2008, the ASCE estimated that a five-year investment of $1.6 trillion is needed to bring the infrastructure to a higher rating. Part of the problem is the lack of investment by government agencies. Money is scarce for infrastructure projects. Clearly there will be pressure on the construction industry to innovate and complete infrastructure projects within budget and on time.

The Importance of Ethics in the Construction Industry

Ethics is an important issue in the construction industry and will be discussed in several chapters of this book. Ethical behavior is a behavior that can philosophically be judged to be "good" or "right." Construction contractors have a responsibility to behave in an ethical way toward their clients and the greater public. Unethical behavior in the construction industry occurs in different ways. It can include unethical and collusive bidding practices, as well as low-quality construction that does not meet quality standards. Improper and illegal bidding practices add to the costs of projects and can reduce quality. The costs of poor quality to the owner and public can be large because substandard facilities will need to be reconstructed sooner then their anticipated design life. Additionally, in the construction industry, "cutting corners" can have hazardous and even fatal consequences to the public.

A good example showing the importance of ethics is described in the following case in Hong Kong. From 1998 to 2001, over 15 construction projects were found to have used shortened foundation piles. The foundation contractors had paid bribes to the engineers employed by the main contractor to produce fake records. The worst case was found in a housing project where two newly completed 34-story skyscrapers had to be demolished because they were structurally unstable and could not be repaired. The cost to the tax-paying public was HK$600,000,000 (Ho et al., 2004). This example illustrates how unethical behavior can cost society vast sums of money, and endanger people's lives.

Careers in Construction

A career in construction can be fascinating and financially rewarding. People choose a career in construction for many reasons. These include the following:

- The ability to work outdoors in a dynamic environment
- The satisfaction of contributing to and seeing an important project evolve from inception to final completion
- Better compensation for construction managers
- Construction can be a good career choice both for people who want to start their own businesses and for people who want to work in an organization. The

majority of homebuilders are small companies. Heavy and industrial construction is typically conducted by very large firms, and most people prefer to work for these firms as employees. However, these large firms offer employees the chance to travel to work on projects all over the world and the opportunity to work on exciting and complex megaprojects.

There are many different educational paths available for people with an interest in construction. There are many undergraduate programs in construction technology, building construction, construction management, and civil engineering that provide a good preparation for a career in construction. People with an interest in residential construction often major in building construction. Those with an interest in heavy or industrial construction may find a civil engineering degree to be good preparation for a career as a construction manager.

Summary

This chapter has provided an introduction to the construction industry. Important points to consider are the following:

1. Construction is an important segment of the U.S. economy employing millions of people. Vast sums are spent annually on construction.
2. Construction differs from manufacturing in its emphasis on unique facilities, custom-built on a project basis.
3. There are many stakeholders in construction projects including the owner, the designer, the construction contractor, and the community adjacent to the project location.
4. There has been an increasing trend to build complex, multiyear megaprojects.
5. The construction industry is becoming global in nature, requiring construction contractors to understand the culture and business practices of many countries.
6. Ethics is an important consideration for everyone in the construction industry. Ethical behavior is required of all parties in construction to construct high-quality projects at the lowest cost.
7. The construction industry offers exciting careers for people wishing to start a small company or those wishing to work on complex projects for a large contractor. Working in the dynamic environment of construction can be truly rewarding.

Key Terms

Brownfield projects
Construction contractor
Designer
Globalization
Greenfield projects
Gross domestic product (GDP)
Infrastructure
Megaprojects
Owner
Project
Retrofitting
Subcontractor

Review Questions

1. What are the characteristics of a megaproject? Why are megaprojects often controversial?
2. Describe the role of the owner, designer, and contractor in the construction process.
3. How is the construction industry different from a manufacturing industry?
4. Who are the major participants in a construction project?
5. Describe ways a construction project in a foreign country can be riskier than a domestic project.
6. What do you think would be the consequences of unethical behavior for a construction professional?
7. What types of construction interest you the most? Do you want to start your own company or work as an employee for a large construction company?

Management Pro

MANAGEMENT PRO

Identify a megaproject taking place in your region. What is the project cost? How long is its completion intended to take? What are the characteristics that make it complex and different from smaller projects? Is there any controversy related to the project? What special management skills does the project require? Write a short report defining the major challenges of the project.

References

American Society of Civil Engineers. Report card for America's infrastructure. 2005. Available from http://www.asce.org/reportcard/2005/index.cfm (accessed January 21, 2009).

Bauml, Steven. 1997. Engineering and construction: Building a stronger global industry. *Journal of Management in Engineering* 13 (5): 21–24.

Ching, Francis D.K. and Steven R. Winkel. 2007. *Building Codes Illustrated: A Guide to Understanding the 2006 International Building Code*, 2nd ed. Hoboken, NJ: John Wiley.

Flyvbjerg, Bent, Nils Bruzelius, and Werner Rothengatter. 2003. *Megaprojects and Risk: An Anatomy of Ambition*. Cambridge, UK: Cambridge University Press.

Ho, Man-Fong, Derek Drew, Denny McGeorge, and Martin Loosemore. 2004. Implementing corporate ethics management and its comparison with the safety management system: A case study in Hong Kong. *Construction Management and Economics* 22 (July): 595–606.

Lindberg, Brian M. and Justin M. Monaldo. 2008. Annual industry accounts: Advance statistics on GDP by industry for 2007. Washington, D.C.: U.S. Bureau of Economic Analysis. Available from http://www.bea.gov/scb/pdf/2008/05%20May/0508_indy_acct.pdf (accessed January 21, 2009).

Nakheel L.L.C. 2006. The Palm. Available from http://www.nakheel.com/Developments/The_Palm/ (accessed November 15, 2006).

Ngowi, A.B., E. Pienaar, A. Talukhaba, and J. Mbachu. 2005. The globalization of the construction industry—A review. *Building and Environment* 40: 135–141.

Project Management Institute. 2004. *A Guide to the Project Management Body of Knowledge*, 3rd ed. Newtown Square, PA: Project Management Institute.

The San Francisco–Oakland Bay Bridge Seismic Safety Project. 2006. Corridor overview. Available from http://www.baybridgeinfo.org/Display.aspx?ID=8 (accessed December 11, 2006).

U.S. Census Bureau. 2009a. The 2009 statistical abstract, 921—Value of new construction put in place. Available from http://www.census.gov/compendia/statab/tables/09s0921.pdf (accessed January 21, 2009).

U.S. Census Bureau. 2009b. The 2009 statistical abstract, 928—New privately-owned housing units started by state. Available from http://www.census.gov/compendia/statab/tables/09s0928.pdf (accessed January 21, 2009).

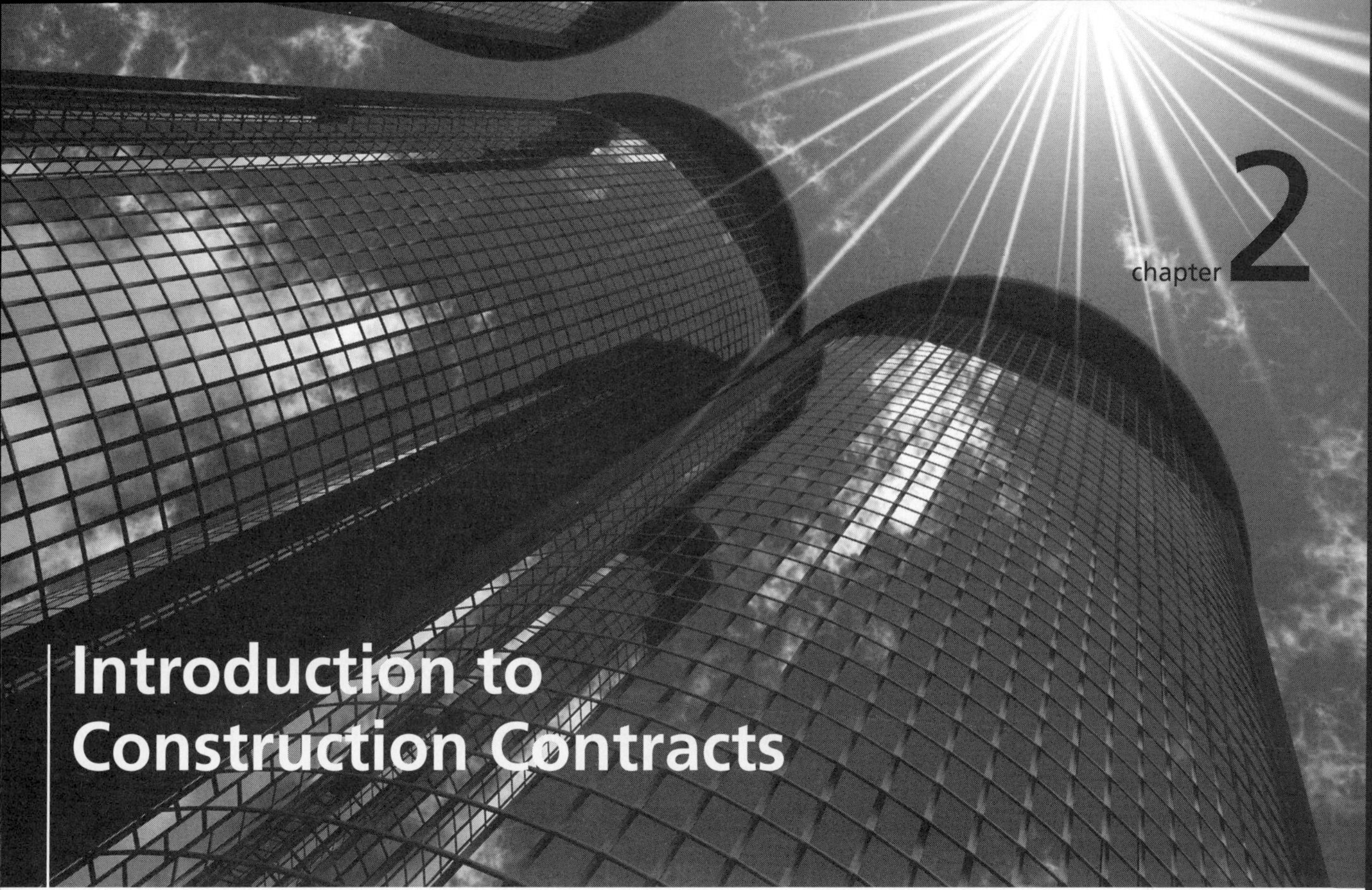

chapter 2

Introduction to Construction Contracts

Chapter Outline

Introduction

In the modern world, written **contracts** are used to define the legal responsibilities of the parties involved in an agreement. A contract is an exchange of promises between two or more parties to commit to actions that are enforceable in a court of law. In construction, the written and graphical documents produced by the designer become a legal contract between the owner and contractor, defining exactly what is to be constructed. For a construction project, the exchange of promises occurs when the owner promises funds to pay for a completed facility and in exchange the contractor promises to produce a completed facility. As will be illustrated in this and the following chapters, construction contracts can be the source of many disputes between the owner, contractor, and designer. Often, the owner and the contractor may have differing interpretations of the design, which lead to increases in project cost, claims, and costly litigation.

In this chapter, we will explore some of the construction contractual arrangements that are most common. We will discuss the steps in the competitive bidding process in detail. We will define and explain the various documents that are typically included in a construction contract. We will also discuss contract modifications that occur during construction.

Basic Construction Contractual Arrangements

Figure 2.1 illustrates the traditional arrangement of contracts for a construction project. In this arrangement, there are two primary contracts. These are:

1. **The contract between the designer and the owner.** The owner pays the designer to produce plans and specifications that describe what is to be built. Depending on the type of project to be constructed, the lead designer may be an architect or an engineer. Due to the complexity of many projects, the lead designer may hire other firms to complete parts of the design. For example, an architect may be the lead designer for a skyscraper. The architect will define the form of the building and its aesthetics and hire specialists like a structural engineer to decide on various parameters of the building structure, such as the dimensions of beams and columns to be used. A mechanical engineering firm that specializes in heating, ventilation, and air conditioning systems may also be hired.
2. **The contract between the owner and the construction contractor.** This contract obligates the construction contractor to build the project in return for payment by the owner. The contractor must build in accordance with the plans and specifications produced by the designer. However, it is important to note that in this arrangement, there is no contractual arrangement between the designer and the contractor, although they must interact significantly on any project.

   The construction contractor who is awarded the project often functions as a "prime" contractor. The prime contractor subcontracts portions of the work to specialty contractors. The subcontractors have a contract with the prime contractor but no direct contractual relationship with the owner.

Figure 2.1

Traditional contractual arrangements

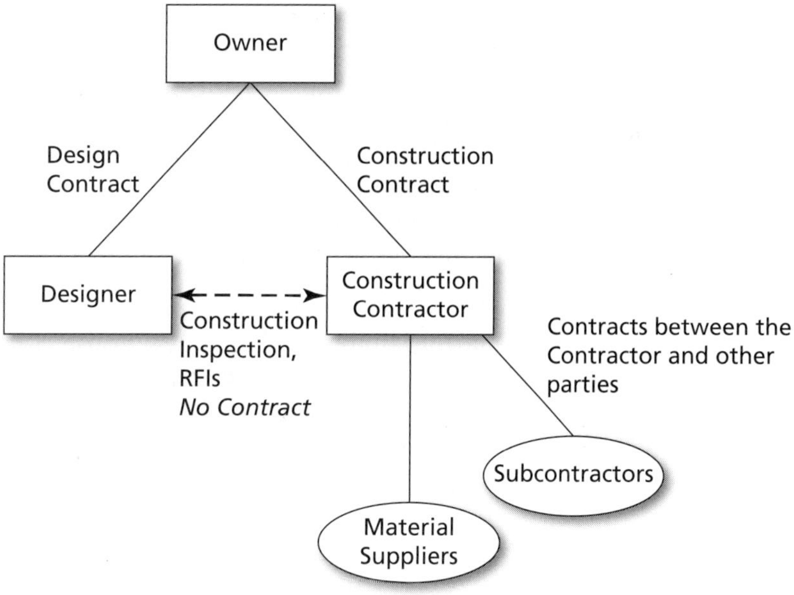

Public and Private Owners

At this juncture, it is prudent to consider how the owner selects a construction contractor and a designer. In some circumstances, the methods of selection of the designer and the contractor are completely different. Designers are almost always selected based on merit, that is, a design firm is selected based on the firm's capabilities and qualifications. This is sometimes called a quality-based selection process. The steps in a quality-based selection process can be described as follows:

- The owner issues a request for qualifications (RFQ). Interested designers submit their qualifications in response to the RFQ. Qualifications submitted by the design firms are a description of the firm's capabilities and information about successful projects of a similar nature they have designed.
- The owner prepares and selects a short list of the most qualified designers. In some instances, the owner may meet with the short-listed designers to better assess their qualifications.
- The owner ranks and selects who he or she believes is the most qualified designer.
- A scope of work and fees are negotiated between the owner and designer, who negotiate on exactly what is to be designed. The amount the designer is to be paid is also determined through negotiation.

Using this method, the owner selects designers based on their merits, and not by the designer who submits the lowest fee. This is considered the best way to insure that the design is of the highest quality possible.

When selecting a construction contractor, there has traditionally been a clear demarcation between publicly and privately funded projects. Publicly funded construction projects are typically initiated by a government agency and are funded using taxpayer money or bond issues. Examples of publicly funded projects are the construction of a new highway by a state department of transportation and construction of a new sewage treatment plant in a city. To insure fairness in awarding contracts, government agencies must follow many laws and regulations. In particular, the construction contractor for public construction projects is often awarded the contract using a process of competitive bidding.

In privately funded construction, the money for the project comes from many sources. It can include a family building a new house, a large corporation expanding a factory, or a real-estate developer constructing a new shopping mall. The key difference in this type of construction is that a private owner is not bound by the procurement laws and regulations that require government agencies to use competitive bidding. Private owners can hire contractors through negotiation or through competitive bidding.

Construction Contract Types

There are several different contract types that can be used in construction. The differences are based on the way the contractor is compensated and on the way risk is allocated between the owner and the contractor. Risk can be defined as the potential for financial loss. The four types of contracts most frequently seen are:

- Lump sum
- Unit price
- Cost-plus-fee
- Guaranteed maximum price

Competitively bid projects are typically conducted using lump sum or unit price methods. Negotiated projects in the private sector are often conducted using cost-plus-fee or guaranteed maximum price contracts.

Lump sum contracts are most frequently seen in building and residential construction. The contractor offers to build the facility for a single price. This fixed price of the contract is difficult to change and does not handle variations in quantity very well. This type of contract is considered to be riskier for the contractor than the owner.

Unit price contracts are more often seen in heavy construction because it is often difficult to develop exact estimates for earthwork and subsurface work. In this type of contract, the project is divided into many separate pay items. For each pay item, the contractor must develop a cost per unit. For example, pavement stripes may be paid in dollars per linear foot. The contractor would know from guide quantities provided by the owner that there are 1,000 linear feet of stripes in a project. If the contractor decided to bid $2/linear foot, the total cost for the striping pay item would be $2,000. This type of contract is less risky for the contractor because variations in quantities can be adequately handled. If it turned out that 1,075 linear feet of stripes were actually required, the contractor would receive $2,150. The unit price contract is riskier for the owner because the final project cost cannot be firmly established, and if the guide quantities provided have been underestimated, the project could be considerably more costly. Most general conditions include provisions to renegotiate unit prices if there are significant deviations between estimated and actual quantities.

In the cost-plus-fee contract, the contractor is reimbursed for all costs plus a fixed fee that has been negotiated between the owner and contractor. This type of contract is less risky for the contractor because he or she will be reimbursed for all costs. It is riskier for the owner because there is no maximum price fixed for the construction project. This has led to a more widespread use of the guaranteed maximum price (GMP) contract. In this type of contract, the contractor is reimbursed for costs up to the GMP. If the contractor runs over, he or she must cover the additional costs.

Competitive Bidding

In competitive bidding, construction contractors submit a price for the construction of a project, and the contractor with the lowest bid is selected to construct the project. Competitive bidding is believed to be the fairest method of awarding public projects. However, it has some disadvantages. Foremost among them is the requirement for the design to be complete before construction can commence. The bidders need a completed design to prepare their bid. Therefore, it is very difficult to compress the amount of time needed to construct the project. Figure 2.2 shows the time period involved for a competitively bid project. First, there is the design stage, which may last for a substantial period of time for a complex project, followed by a bidding period where bidders study the design and calculate their prices; finally, after the bid opening, construction can commence.

Figure 2.2

Phases of a competitively bid project

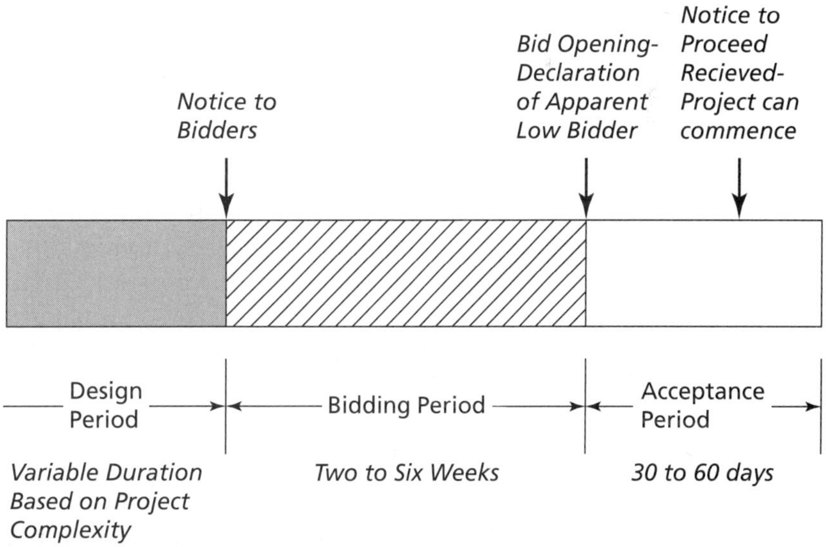

Unit Price and Lump Sum Projects

There are two types of competitive bidding, lump sum and unit price. In lump sum projects, the contractor provides a single price for the entire project. In unit price competitive bidding, the project is broken into various line items. Contractors bidding on a unit price project develop a cost per unit for each line item. Bids are opened at the end of the bidding period and the bidder with the lowest price wins the project.

Steps in the Competitive Bidding Process

The bidding process consists of a set of steps that span the period from the completion of the design to the award of contract. Construction contractors must receive the completed design to use in the preparation of their bid estimate.

Completion of the Design and the Development of the Bid Package

Using competitive bidding, the design must be 100% complete before bidding can commence. The designer will have produced a complete set of plans and specifications. Additional documents will be added to these core documents to form a **bid package.** The bid package consists of all the documents that contractors must consider when deciding to bid on a project. The bid package documents will become legally binding on the low bidder and will be incorporated in the construction contract between the owner and the low bidder.

The **plans** and **specifications** describe the details of the design. The plans are a graphical representation of what is to be built. The specifications are a textual description of general information and quality requirements of items shown in the plans. The documents added to the bid package include the **proposal form, general conditions,** and **special conditions.** The proposal form is the document used by the contractor to specify his or her bid prices. The general conditions and special conditions are important documents that will be discussed in detail in Chapter 3. They set forth the contractual relationships between parties and describe how the project is to be administered.

Advertising the Bid

The primary concern of the owner is to insure that there is a lively competition that results in the bidding contractors submitting reasonable market prices to complete the project. To insure competition, the owner advertises the availability of the project. Advertising will make sure that the greatest numbers of potential bidders hear about the project. Traditionally, a notice to bid was published by the owner in newspapers and trade magazines. However, Web-based systems are now being used by many owner organizations to supplement or supplant the printed announcements.

Obtaining the Bid Package

When the bid package is complete, construction contractors must obtain the bid package. In the past, for complex projects with many plan sheets and thick specifications, owners often charged contractors several hundred dollars to purchase the documents. In some areas, plan rooms that provide a central location were available where contractors could go to examine plans and specifications for building projects. In the recent past, online plan rooms have become popular. These services allow contractors to download files, typically in the Adobe PDF format. Some of these online services are free whereas others require a subscription. This is the first stage where a contractor must make a decision about how a project "fits" with his or her firm's capabilities. Preparing a bid estimate can be a costly undertaking, depending on project complexity. If contractors do not believe they have adequate expertise or there is a low chance of being the winning bidder, they will not proceed further with studying the project. If the contractors believe they have a reasonable chance of being the low bidder, they will obtain the plans and proceed with preparing cost estimates for the project.

Preparing the Bid

We have already discussed the proliferation of megaprojects (see Chapter 1). With increasing project complexity, the time and effort required to prepare a bid estimate have increased. Preparing a successful bid estimate, however, is vital to the success of the construction company. A successful construction company must be able to produce winning bids for a reasonable number of projects, yet make sure that the bid amount is high enough for the company to be profitable. A successful construction bid must be comprehensive to include:

1. **The cost of construction.** The bid must cover the direct costs of the labor, materials, and equipment required to build the project.
2. **Project and home office overhead.** The bid price must include the cost to the contractor for supervision of the construction project and the cost of maintaining an office that includes the administrative functions of the construction company.
3. **The contractor's profit.** Importantly, the bid must include a profit for the contractor. The determination of the amount of the profit is a complex consideration. Naturally, a bidder would want to make as much profit as possible. However, including a large profit in a bid may make it uncompetitive. Therefore, bidders must balance the profit included in the bid with the need to provide a price that is competitive. Many external factors must be considered, including the extent of competition, the bidders' need to get new work, and the general state of the economy.

Design Changes During the Bidding Period

Design changes are often made during the bidding period and are called **addenda**. It is important for the bidding contractor to keep track of the changes in the design and make sure his or her bid reflects all the addenda that have been issued. The addenda that are issued legally become part of the bid package.

Pre-Bid Meetings

Most projects have a meeting where the owner and designer are available to answer questions about the proposed project. It is very important for the contractor to attend this meeting to insure that he or she has not misinterpreted the project requirements. Attendance at this meeting can forestall potential bidding mistakes.

Submitting the Bid and the Bid Opening

Bids may be submitted until the time of the **bid opening**. At the bid opening, all the bids are opened and the low bidder is declared. Traditionally, bids are submitted to a bid box in a location designated by the owner. Typically, construction bids are submitted at the last moment because of the need to collect many price quotations of subcontractors and material suppliers for different elements of the project. Before the time of the bid opening, submitted bids may be withdrawn. No bids may be submitted after the time of the bid opening, nor may bids that have already been submitted be modified.

In public contracting, projects are typically awarded to the contractor with the lowest responsible and responsive bid. The lowest bid is not always awarded the contract, and owners often reserve the right to reject any and all bids. A responsible bidder is capable of undertaking the project and completing it in a satisfactory manner. When determining if a contractor is responsible, the contractor's financial resources, performance record, integrity, and technical capabilities are assessed (Kelleher, 2005). Sometimes, pre-award meetings are held by the owner with the low bidder to learn the contractor's strategy for the project and to identify potential problems with the bid. If the owner on the basis of this meeting believes the low bidder is not responsible, the low bidder may be declared so. Some government agencies **prequalify** bidders before they can be eligible to receive a bid package. The owner agency conducts a responsibility inquiry of the financial, managerial, and performance records of prospective bidders. Only bidders that the agency prequalifies are allowed to bid on the project (Jervis and Levin, 1988).

A responsive bid is an unqualified offer to perform the work in exact accordance with the plans and specifications. Sometimes, a bidder may try to deviate in some way from the terms of the bid solicitation, such as stating that they reserve the right to modify bid prices if cost increases occur. A bid that deviates in this way would be declared nonresponsive and would be rejected by the owner.

Award of Contract and Notice to Proceed

After the low bidder has been determined to be responsible and responsive, the contract is awarded and a formal agreement signed. This agreement includes wording that incorporates the bid package documents formally into the contract agreement. The project commences when a written **notice to proceed** is sent by the owner to the contractor. The notice to proceed indicates to the contractor that the project site is available and that he or she may commence work. It establishes the starting date of the project.

Bid Security and Bid Bonds

As part of a bid submission, an owner will normally require contractors to obtain a **bid bond** to provide bid security. This is done to protect the owner from financial loss if the low bidder cannot construct the project. A bid bond is a form of insurance that is purchased by the contractor from a surety company. The bid bond assures the owner that if the low bidder cannot enter into a contract, then the owner will not be responsible for the difference in cost between the original low bidder's price and the second lowest bidder's price. The surety will step in and pay the owner the difference (Jackson, 2004).

Alternatively, another solution that is sometimes seen in the construction industry is for a contractor to post actual funds as security against failure to perform the project. A bid security would call for a contractor to submit a certified check for 10% of the bid amount. If the contractor is the low bidder and then backs out of the project, the contractor loses the money. Otherwise, the bid security check is returned to the bidders.

Bid Package Details

In this section, we will discuss the details of the documents that are included in the bid package. A successful construction contractor must understand the details and legal ramifications of these documents.

Specifications

Specifications are an important part of the bid package. Specifications can be thought of as detailed instructions to the builder. They provide a written description of the work to be performed and the materials to be installed.

Professional organizations and large owners who initiate many construction projects have developed standard specifications. In particular, for highway construction, many state DOTs have extensive and detailed standard specifications for items that are repeated frequently on their projects. The Army, Navy, and NASA have also published extensive specifications. Figures 2.3 and 2.4 show a portion of the standard specification of the Western Federal Lands Highway Division. This is an organization that is part of the U.S. Department of Transportation and is responsible for highway construction on federal lands such as national parks. The specification illustrated in the figures is for "clearing and grubbing." This is the construction activity where ground is cleared and stumps removed in the first steps of building a new road. Note that this specification is numbered. Most agencies apply a standard numbering system.

The clearing and grubbing specification is broken into several sections that describe the materials to be used: the construction requirements, the method of measurement of the work performed, and the method of payment. Specifications can be complex and will often reference other standard specifications. In Figure 2.3, the specifications defining material requirements are circled. Included in the circled materials is backfill. Figure 2.4 shows the specification that gives requirements for the backfill material that is to be provided.

Section 201

Section 201. — CLEARING AND GRUBBING

Description

201.01 This work consists of clearing and grubbing within the clearing limits designated on the plans.

Material

Specifications defining material requirements

201.02 Conform to the following Subsections:

Backfill material	704.03
Tree wound dressing	713.05(g)

Construction Requirements

201.03 **General.** Construct erosion control measures according to Section 157. Perform work within designated limits. Do not damage vegetation designated to remain. If vegetation designated to remain is damaged or destroyed, repair or replace the vegetation in an acceptable manner. Where possible, preserve all vegetation adjacent to bodies of water. Treat cuts or scanned surfaces of trees and shrubs with tree wound dressing.

201.04 **Clearing.** Within the clearing limits, clear trees, brush, downed timber, and other vegetation as follows:

(a) Cut all trees so they fall within the clearing limits.

(b) In areas of cut slope rounding, cut stumps flush with or below the final groundline.

(c) In areas outside the excavation, embankment, and slope rounding limits, cut stumps to within 6 inches of the ground.

(d) Trim trees branches that extend over the road surface and shoulders to attain a clear height of 20 feet. If required, remove other branches to present a balanced appearance. Trim according to accepted tree surgery practices. Treat wounds with tree wound dressing.

201.05 **Grubbing.** Grub deep enough to remove stumps, roots, buried logs, moss, turf, or other vegetative debris as follows:

(a) Grub all areas to excavated except for cut slope rounding areas.

86

Figure 2.3 Clearing and grubbing specification

Section 201

(b) **Grub al embankment areas.** Undisturbed stumps maybe left in place if they protrude less than 6 inches above the original ground and will be covered with more than 4 feet of embankment.

(c) Grub pits, channel changes, and ditches only to the depth necessary for the excavation.

(d) Backfill stump holes and other grubbing holes with backfill material to the level of the surrounding ground according to Subsection 209.10. Compact backfill according to Subsection 209.11.

201.06 **Disposal.** Merchantable timber in the Contractor's property. Dispose of clearing and grubbing debris according to Subsection 203.05.

201.07 **Acceptance.** Clearing and grubbing will be evaluated under Subsection 105.02.

Material for tree wound dressing will be evaluated under Subsection 106.03.

Backfilling and compacting of stumps and grubbing holes will be evaluated under Section 209.

Measurement

201.08 Measure the Section 201 items listed in the bid schedule according to Subsection 109.02 and the following as applicable.

Do not make deductions from the area computation unless excluded areas are identified in the contract.

Do not measure cleaning and grubbing or borrow or material sources.

Payment

201.09 The accepted quantities will be paid at the contract price per unit of measurement for the Section 201 pay items listed in the bid schedule. Payment will be full compensation for the work prescribed in this Section. See Subsection 109.05.

87

Figure 2.3 *Continued*

Figure 2.4

Reference to the AASHTO standard in specification

704.03 **Backfill Material.** Furnish a well-graded, compactable material free of excess moisture, mock, frozen humps, roots, sod, or other deleterious material conforming to the following:

(a) For all structures and pipes other than plastic pipe:

(1) Maximum particle size	3 inches
(2) Soil classification AASHTO M 145	A-1, A-2, or A-3

(b) For plastic pipe:

(1) Maximum particle size	1½ inches
(2) Soil classification, AASHTOM 145	A-1, A-2-4, A-2-5, or A-3

Figure 2.4 also illustrates the use of testing standards in specifications. AASHTO M 145 has been circled in the specification. What this tells the construction contractor is that the determination of the soil classification for the backfill material must conform to soil testing procedures that have been established by the American Association of State Highway and Transportation Officials.

Figure 2.5 shows a specification for electrical work for highway signals and lighting. The circled area in this specification shows that the work must conform to specifications in the National Electrical Code and to specifications from other national organizations. This shows that specifications will often incorporate work requirements from other sources. These examples illustrate that a contractor must be aware of both standard material testing requirements and standard construction requirements promulgated by national organizations.

In commercial building construction, the Construction Specifications Institute (CSI) has authored MasterFormat, which is an indexing system for organizing construction specifications. The MasterFormat is a list of standardized numbers and titles for organizing construction specifications by work results. A work result is something like "painting" or "lighting." The purpose of the MasterFormat is to provide a framework for organizing a complex set of specifications. The highest level of organization is groups, subgroups, and divisions. For example, under the specifications group, there is a facility construction subgroup that is further divided into 19 divisions. Example divisions include concrete, metals, finishes, and equipment. A CSI number typically consists of six digits. The first pair defines the division, the second pair defines the broad scope, and the third pair defines a medium scope. Additional numbers can be added for greater specificity. The number is associated with a title which is a work result.

Standard prewritten specifications conforming to CSI formats and methods are available from commercial master guide specification systems. These systems feature computer software that allow a user to access a database of text specifications,

Section 201

Section 636. — SIGNAL, LIGHTING, AND ELECTRICAL SYSTEMS

Description

636.01. This work consists of installing, modifying or removing traffic signals, flashing beacons, highway lighting, sign illumination, communication conducts, and electrical systems or provisions for future systems.

Material

636.02 Conform to the following Subsections:

Backer rod	712.01(g)
Electrical material	721.01
Lighting material	721.02
Precast concrete units	725.11(d)
Sealant	712.01(a)

Construction Requirements

6360.03 **Regulations and Codes.** Furnish material and workmanship conforming to the standards of the National Electrical Code, local safety code, UL, and National Electrical Manufacturers Association.

Obtain permits, arrange for inspections, and pay all fees necessary to obtain electrical service.

Furnish luminaries with crashworthy supports.

Notify the CO, local traffic enforcement agency, local utility company, or railroad company 7 days before any operational shutdown to coordinate connections or disconnections to an existing utility or system.

636.04 **General.** At the preconstruction conference, submit a certified cost breakdown of items involved in the hump sum for use in making progress payments and price adjustments.

Fifteen days before installation, submit a list of proposed equipment and material. Include the manufacturer's name, size, and identification number of each item. Supplement the list with scale drawings, catalog cuts, and wiring diagrams showing locations and details of equipment and wiring.

154

Figure 2.5 Reference to the building code in specification

and then modify the prewritten specification according to the specific requirements of a project. Examples of commercially available systems include MasterSpec (www.arcomnet.com), SpecText (www.spectext.com), and BSD SpecLink (www.bsdsoftlink.com).

Some federal government agencies, including the Army, Navy, and NASA use a computer system called SpecsIntact (SpecsIntact, 2007), which is used for producing specifications for government construction projects. The SpecsIntact system uses a database of master guide specifications created by the agencies. Using the guide specification, the users adopt them to their project. The specifications are divided into divisions according to the CSI format. Figure 2.6 shows two windows from the SpecsIntact system. The upper window shows a sample project with the CSI specification items that have been selected for inclusion. The lower window shows the actual text for the masonry specification. The text includes blanks where the designer can fill in information appropriate for the

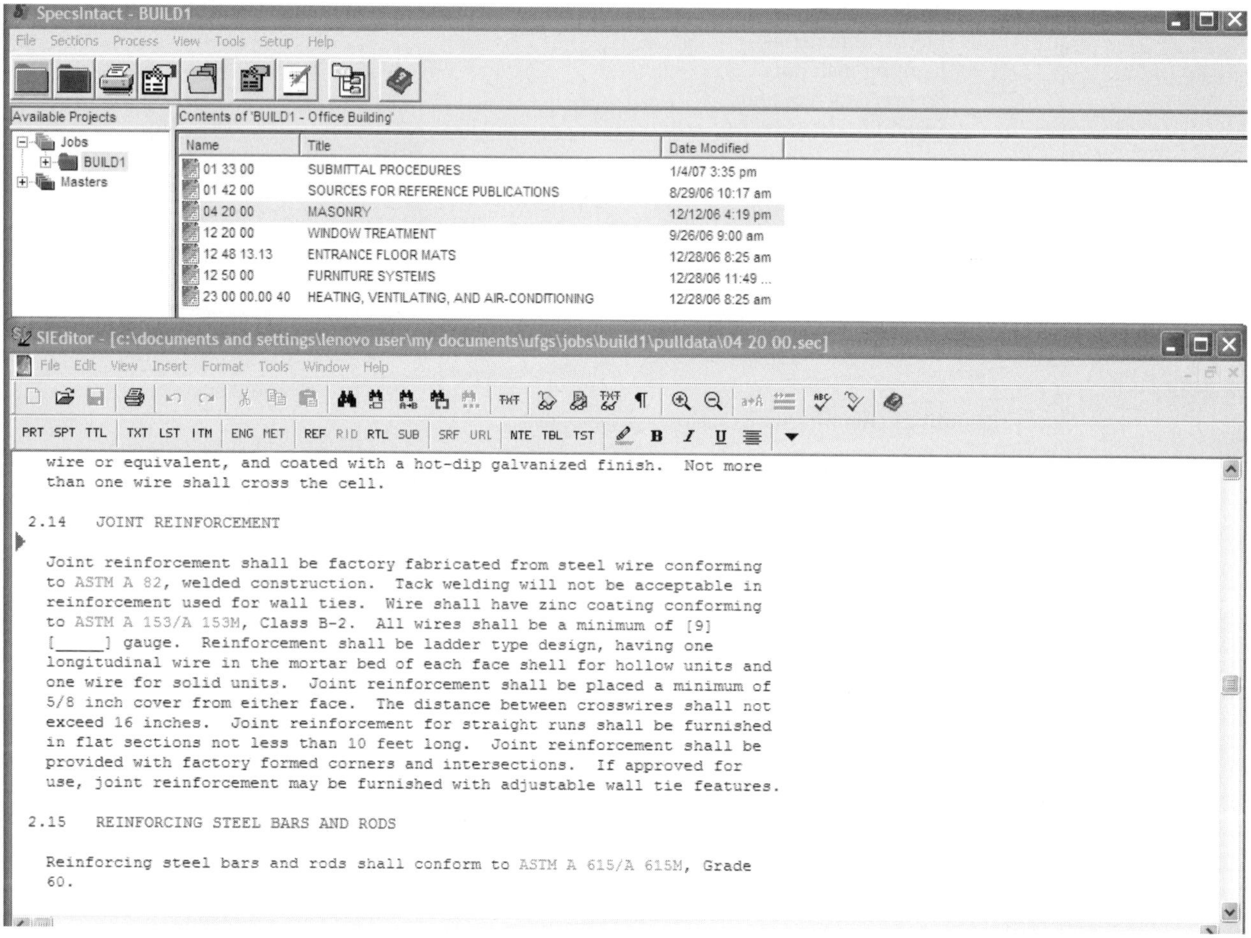

Figure 2.6 A sample project specification using the SpecsIntact system

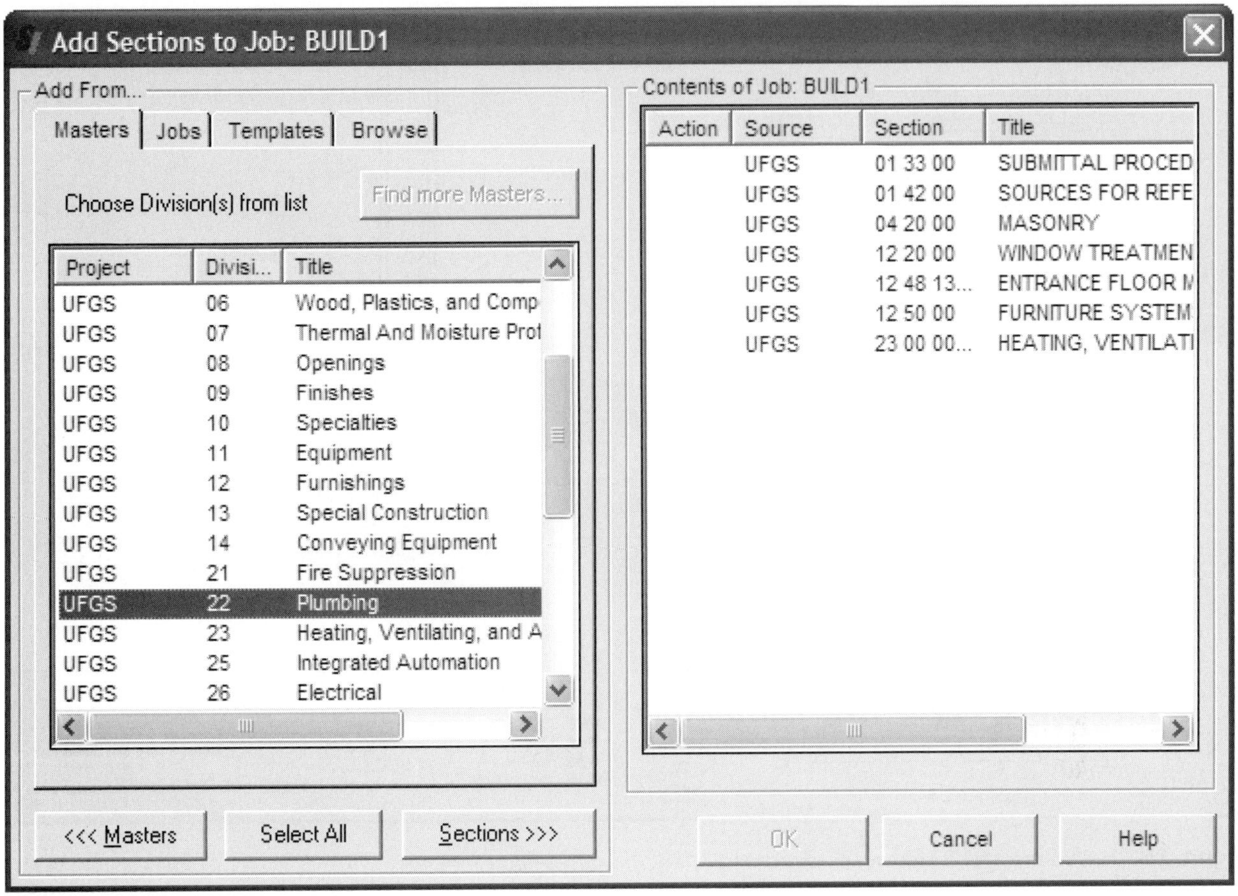

Figure 2.7 Adding additional specifications using SpecsIntact

project. The designer can also modify the specification if necessary. Figure 2.7 shows how additional specifications are added to a project. A user selects from the master specification listing and transfers the needed specifications to a project. The SpecsIntact program has the capability to automatically publish a specification as a PDF or Word file. More information about SpecsIntact can be found at specsintact.ksc.nasa.gov/.

Project Plans

Plans are a graphical description of what must be constructed. Figure 2.8 shows a plan sheet from a typical highway construction project. This particular plan sheet shows a landscaping plan for a new replacement bridge and traffic circle. Figure 2.9 shows a general plan and elevation for a bridge that is to be constructed as part of this project. A typical project will have many plan sheets describing different aspects of the project. This project required a total of 122 different sheets. The plan sheet

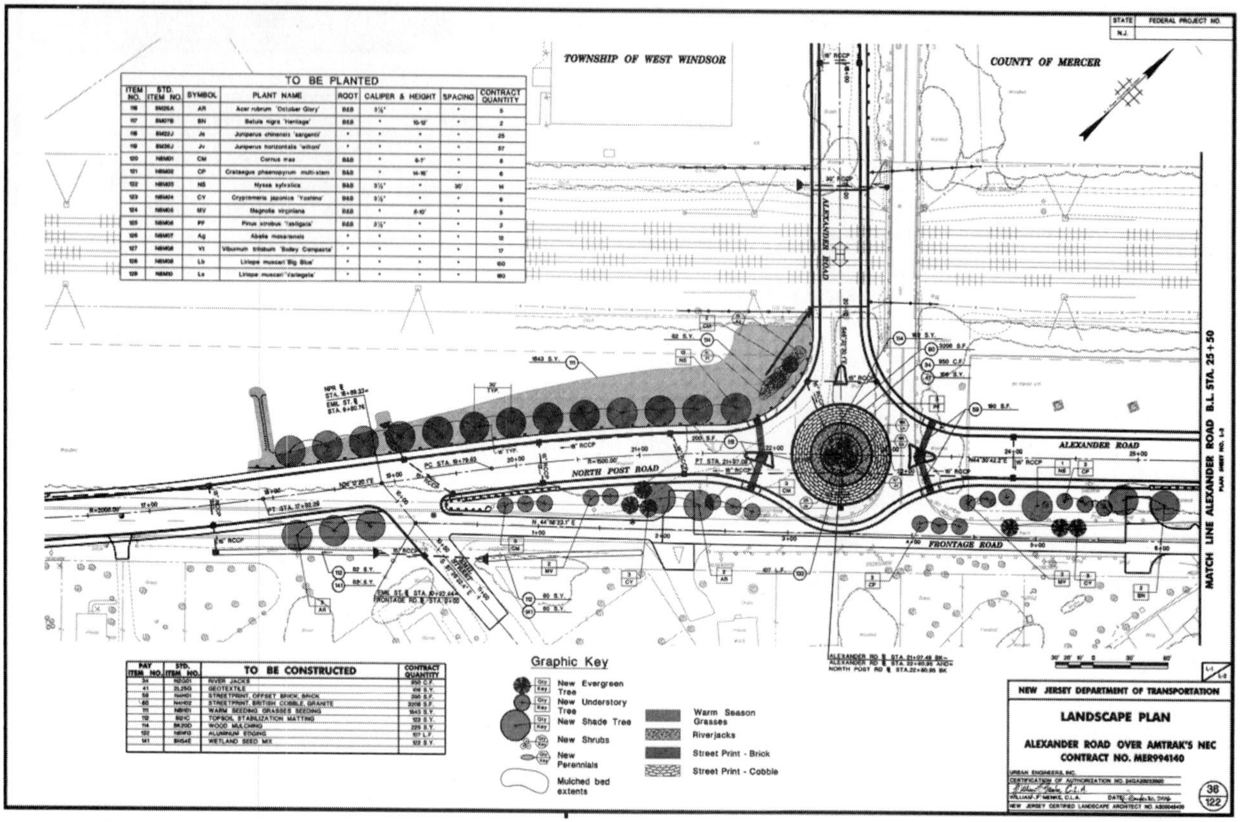

Figure 2.8 Plan sheet for a highway construction project. *Courtesy of Info Tech, Inc.*

shows what is to be constructed and references items that are included in the written specifications.

The Proposal Form

The proposal form is an important document because the contractor provides his or her prices to perform the construction work on this form. For a lump sum bid, only the total project price is submitted for the work. For unit price contracts, a unit price is provided for each project line item. Then, each unit price is multiplied by the guide quantity provided on the proposal form to determine the cost for that line item. The quantities for each unit on the proposal form are determined by the designer. It is always good practice for a contractor to check the quantities provided before bidding to determine if there are significant discrepancies between the guide quantities and the quantities found from the contractor's own analysis of the plans. Finally, the total bid price is established by totaling the price for each line item.

Figure 2.10 shows a proposal form for a New Jersey Department of Transportation construction project. The proposal form has been downloaded from the Web as a

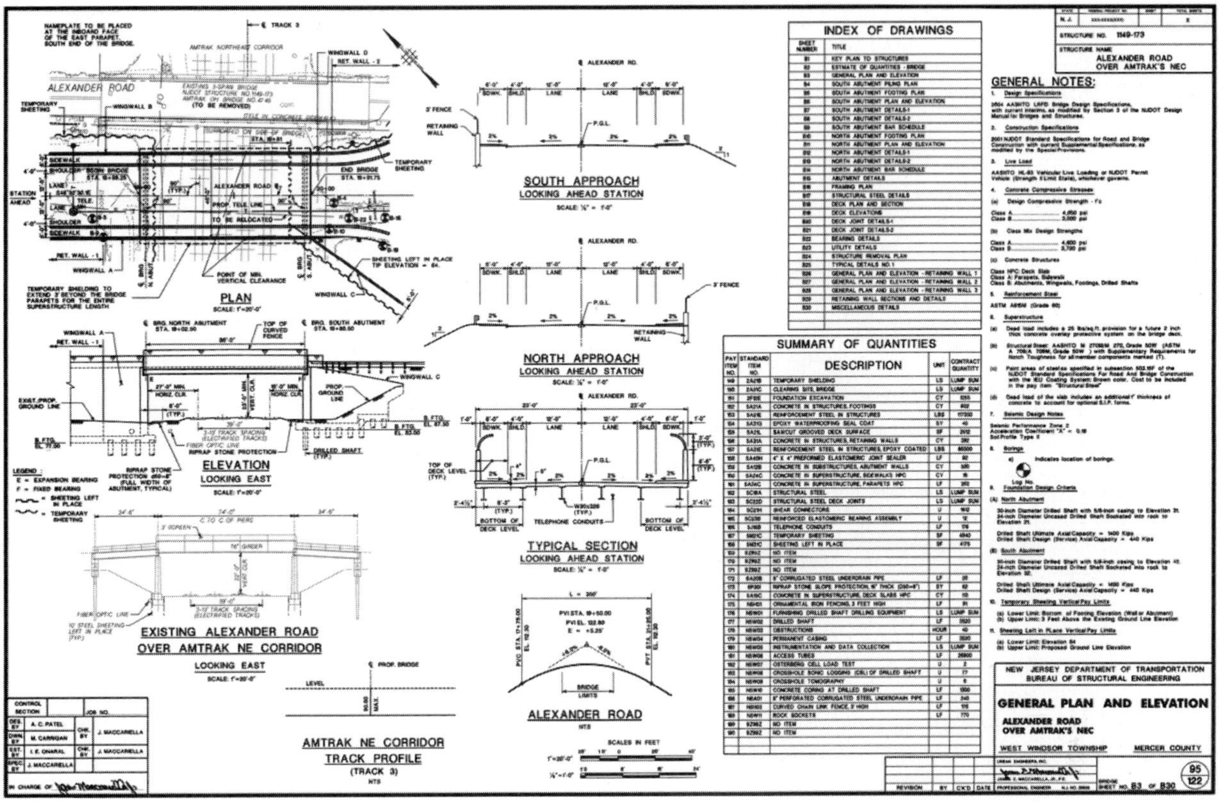

Figure 2.9 A landscaping plan for a new bridge and traffic circle. *Courtesy of Info Tech, Inc.*

computer file, and is being viewed using a program called Expedite. This program allows contractors to submit their bids to state DOTs electronically. The figure shows that the unit's price of $1.25/sf for saw cutting grooved surface deck has been filled in by the contractor and the program has automatically calculated the total price or "extension" for the line item. The numbering of each item corresponds to standard New Jersey DOT specification numbers.

General Conditions

The plans and specifications are the legally binding description of what is to be constructed. The proposal form is where the contractor offers to perform the project for a certain price. Additional documentation must be added to the bid package that describes the contractual relationships between the owner and the contractor. The general conditions are the portions of the bid package where these duties and obligations are defined. Although the language may seem legalistic, a construction contractor must understand all items in the general conditions for any project he or she bids on. Requirements in the general conditions have a direct bearing on contract bid

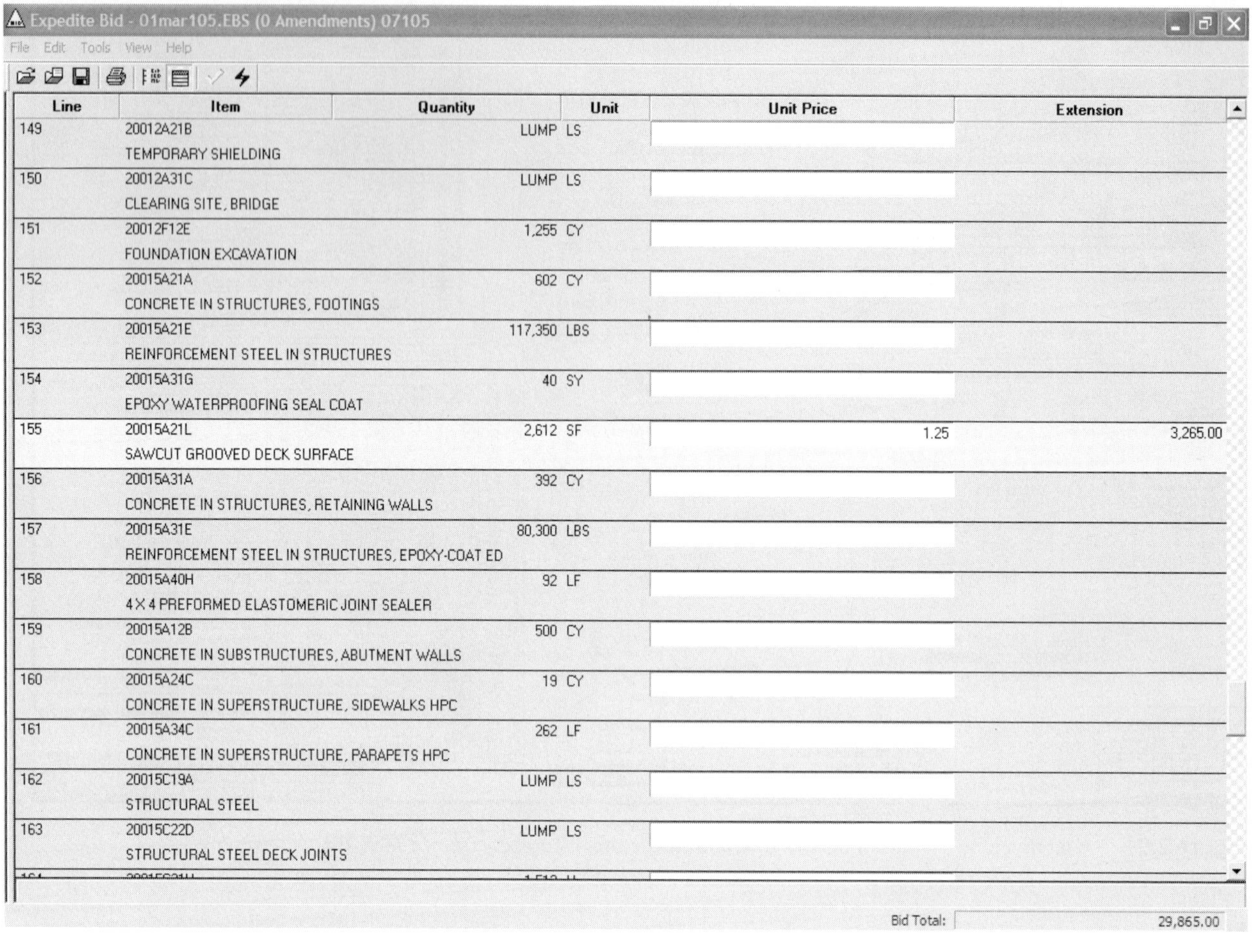

Expedite Bid - 01mar105.EBS (0 Amendments) 07105

File Edit Tools View Help

Line	Item	Quantity	Unit	Unit Price	Extension
149	20012A21B TEMPORARY SHIELDING	LUMP	LS		
150	20012A31C CLEARING SITE, BRIDGE	LUMP	LS		
151	20012F12E FOUNDATION EXCAVATION	1,255	CY		
152	20015A21A CONCRETE IN STRUCTURES, FOOTINGS	602	CY		
153	20015A21E REINFORCEMENT STEEL IN STRUCTURES	117,350	LBS		
154	20015A31G EPOXY WATERPROOFING SEAL COAT	40	SY		
155	20015A21L SAWCUT GROOVED DECK SURFACE	2,612	SF	1.25	3,265.00
156	20015A31A CONCRETE IN STRUCTURES, RETAINING WALLS	392	CY		
157	20015A31E REINFORCEMENT STEEL IN STRUCTURES, EPOXY-COAT ED	80,300	LBS		
158	20015A40H 4 X 4 PREFORMED ELASTOMERIC JOINT SEALER	92	LF		
159	20015A12B CONCRETE IN SUBSTRUCTURES, ABUTMENT WALLS	500	CY		
160	20015A24C CONCRETE IN SUPERSTRUCTURE, SIDEWALKS HPC	19	CY		
161	20015A34C CONCRETE IN SUPERSTRUCTURE, PARAPETS HPC	262	LF		
162	20015C19A STRUCTURAL STEEL	LUMP	LS		
163	20015C22D STRUCTURAL STEEL DECK JOINTS	LUMP	LS		

Bid Total: 29,865.00

Figure 2.10 Expedite proposal form

prices. There are many different topics that are included in the general conditions. They include wording that describes:

- **The duties of the owner and contractor.** The general conditions provide explanations of what is to be provided by the contractor and owner.
- **General project requirements.** Definitions of the project scope and definitions for the meaning of terms used in the contract, plans, and specifications are given in this section.
- **Frequency of payments to the contractor during construction.** Periodic payments to the contractor during construction are typically called **progress payments.**
- **Method of measuring progress and calculating the amount of the progress payment.** How a contractor will be paid and the methods for calculating project progress are described in the general conditions. For example, if the project is done on a lump sum basis, some method will be described that will allow

determination of how much progress has been made in a particular billing period so that the payment the contractor is entitled to can be calculated.

- **Damages for late project completion. Damages or penalties for late completion of a project are described.** Contractors are often assessed monetary **liquidated damages** on a per diem basis for completing a project later than the completion date specified in the construction contract.
- **Mechanisms for extending the project duration.** Projects are often delayed. Often additional time, called a **time extension**, is given to a contractor to complete work if problems beyond his or her control have occurred that delayed the project. The method for requesting a time extension is described, as well as the reasons for which a time extension may be requested.
- **Mechanism for modifying the original design during construction.** Construction projects must often be modified during construction. These modifications often involve a change of cost. These modifications to the project cost are called **change orders** and are negotiated between the owner and contractor. The contractual arrangements governing requests for changes to the original contract are described in the general conditions.
- **Inspection of the contractor's work.** How and by whom the contractor's work will be inspected for compliance with the specifications and quality is defined in the general conditions.
- **Quality requirements.** In the general conditions, sections are typically inserted that give the owner the right to reject work that he or she believes to be of substandard quality.
- **Owner's right to stop or suspend work.** General conditions have clauses that allow the owner to stop or suspend work. This allows an owner's organization to stop a project if it runs out of money.
- **Responsibilities for permits.** The general conditions define who is responsible for obtaining building and environmental permits.
- **Handling of claims. Claims** are unresolved conflicts between the owner and the contractor. General conditions often contain clauses that define how claims are to be handled.
- **Insurance requirements.** General conditions define liability and property damage insurance that must be obtained by the contractor to protect the owner.
- **Bonding requirements.** Several bonds are usually required for a construction project. As we have discussed earlier, a bid bond must be submitted with the bid. The contractor awarded the contract will typically be required to also obtain performance and payment bonds.
- **Definition of retainage requirements. Retainage** is funds withheld from progress payments until the end of the project as an incentive for project completion. The amount withheld is usually 10%. In a large project, this is a substantial amount and provides an incentive for the contractor to finish all work. Retainage is returned with the contractor's last payment. This requirement will be defined in the general conditions.

- **Reasons for contractor termination.** The general conditions define the reasons a contract with a contractor may be terminated before the completion of the project.

Figure 2.11 shows a portion of the general conditions used by the Naval Facilities Engineering Command (NAVFAC) for construction projects for the repair of facilities. This figure shows the general requirements that define the services that the contractor must provide. Figure 2.12 shows a page from the NAVFAC general conditions that describe inspection methods used to establish the quality of the work and procedures to withhold payments from the contractor for unsatisfactory work. A contractor performing a project using these documents would need to be thoroughly familiar with these contract clauses to avoid losing money. A final example from the NAVFAC general conditions is given in Figure 2.13. This figure shows that the contractor is responsible for obtaining all applicable permits to perform the project. This is the type of item that a contractor must be aware of before bidding on a project. There will be costs and manpower that must be expended to get the permits and this must be factored in the bid. In addition, the contractor must consider the amount of time required to obtain the permits so that this activity does not delay the project.

Special Conditions

Most construction contracts will have unique requirements that are not covered by the general conditions. Therefore, unique special conditions are incorporated in

C.1 GENERAL REQUIREMENTS

(1) The Contractor shall furnish all labor, management, supervision, tools, materials, equipment, incidental engineering, and transportation necessary to maintain and repair family housing units and associated utility systems, household equipment, appliances, land areas, and other related real property and facilities in accordance with the contract requirements. Attachment J-C describes the facilities to be maintained in this contract.

(2) The Contractor shall provide services for the following functions: (ex)

PARAGRAPH FUNCTION
C7 Service call work
C8 Preventive maintenance of equipment
C9 Change of occupancy maintenance

a. Regular Working Hours. The Government's regular (normal) working hours are from 0730 to 1630 Mondays through Fridays except (a) federal holidays and (b) other days specifically designated by the Contracting Officer.

(1) Federal Holidays. New Year's Day, Martin Luther Kind, Jr. Day, Presidents Day, Memorial Day, Labor Day, Columbus Day, Veterans Day, Thanksgiving Day, and Christmas Day.

Figure 2.11 Portion of NAVFAC general conditions

E.6 FAC 5252. 246-9303, CONSEQUENCES OF CONTRACTOR'S FAILURE TO PERFORM REQUIRED SERVICES (MAR 1996)

****

NOTE: Insert the following clause in all firm fixed-price and firm fixed-price/indefinite quantity solicitations and contracts for facilities support services.

****

The Contractor shall perform all the contract requirements. The Government will apply one or more of the surveillance methods mentioned below and will deduct an amount from the Contractor's invoice or otherwise withhold payment for unsatisfactory or nonperformed work. The Government reserves the right to change surveillance methods at any time during the contract without notice to the Contractor.

a. STATISTICALLY EXTRAPOLATED SURVEILLANCE METHOD. The Government may apply a statistically extrapolated surveillance method (Random Sampling for Extrapolated Deductions) to any contract requirement to determine Contractor compliance. The defect rate will then be extrapolated to the monthly population to determine the number of unsatisfactorily performed work occurrences. The monthly population is the total number of work occurrences that are required to be performed during the month.

b. OTHER SURVEILLANCE METHODS. The Government may apply other surveillance methods to determine Contractor compliance. These include, but are not limited to, 100% inspection, random sampling without extrapolated deductions, and planned sampling as primary surveillance methods; and incidental inspections and validated customer complaints as supplemental surveillance methods. When using these surveillance methods, deductions may be taken for all observed defects.

c. PROCEDURES. In the case of unsatisfactory or nonperformed work, the Government:

(1) may give the Contractor written notice of deficiencies prior to deducting for unsatisfactory or nonperformed work and/or assessing liquidated damages. Such written notice shall not be a prerequisite for withholding payment for nonperformed work. The Government may specify, as provided for below, that liquidated damages can be assessed against the Contractor. Such liquidated damages are to compensate the Government for administrative costs and other expenses resulting from the unsatisfactory or nonperformed work.

(2) may, at its option, allow the Contractor an opportunity to reperform the unsatisfactory or nonperformed work, at no additional cost to the Government. In the case of daily work, corrective action must be completed within _____ hours of notice to the Contractor. In the case of other work, corrective action must be completed within _____ hours of

Figure 2.12 General conditions describing inspection methods

the bid package. In particular, the special conditions define the contract duration. Typically, the contract duration is defined in days. The commencement date of a project is the date of the notice to proceed is typically specified as sometime within 15 days of receipt of the notice to proceed. On the day of project commencement, the contractor has the number of days specified in the special conditions to complete the project.

```
C.8.11 Permits
The Contractor shall, without additional expense to the Government,
obtain all appointments, licenses, and permits required for the
prosecution of the work. The Contractor shall comply with all
applicable federal, state, and local laws. Evidence of such permits
and licenses shall be provided to the Contracting Officer before work
commences.
```

Figure 2.13 Contractor responsibility for obtaining permits defined in general conditions

The Complexity of Bidding

Bidding is a complicated process. A contractor's bidding strategy to a large extent determines if a firm will be successful or will fail. Contractors are confronted with the problem of determining the correct bid amount. The contractor must attempt to look into the future and determine a bid amount that covers the costs of constructing the project as well as providing a suitable profit. If a contractor is conservative and consistently bids high, then he or she may not win an adequate number of projects. Alternatively, if a contractor bids too low, then he or she may win a project yet will be subject to what economists call the "winner's curse." The winner's curse is where the contractor wins the project but has bid so low it is impossible to make a normal profit and the project may even be completed at a financial loss to the contractor (Dyer and Kagel, 1996). Therefore, contractors must carefully choose the projects they bid on. To submit a successful and profitable bid, contractors must understand the market they are competing in, including a knowledge of their competitors, an understanding of their own financial condition, and existing economic conditions. Chua and Li (2000) have discussed important external and internal factors that affect a contractor's decision to bid and the bid prices. External factors are project related or beyond the contractor's control. Internal factors are firm-related factors and relate to the firm's management and expertise. Some example external factors that must be considered in preparing a bid are:

- Size of the project
- Technological difficulty
- Site accessibility
- Quality of plans and specifications
- Government regulation
- Availability of other projects

Some example internal factors that a contractor must consider when bidding a project are:

- Experience in constructing projects of that type
- Current workload
- Management expertise

- Required project rate of return
- Financial ability to conduct project
- These factors must be considered by the contractor in the decision to bid on a contract. Additionally, the contractor must consider these factors when considering the appropriate level of profit to include in the bid.

The Importance of the General Conditions During Construction

The contractual arrangements defined in the general conditions are important during the construction phase also. Construction projects are complex and it is commonplace for modifications to be made to the design during construction. Change, defined as any event that results in a modification of the original scope, execution time, or cost of work, is inevitable because of the uniqueness of each construction project. For each change made to a construction project, contractors are entitled to adjustments in the project cost and schedule (Hana et al., 2002). The procedures to modify the plans, specifications, costs, and duration are described in the general conditions. In this section, we will explore some of the mechanisms for modifying the original project bid package during construction.

Change Orders

Change orders are formal modifications of the original design, project duration, or cost. Either the owner or the contractor can request a change order. There are many different situations that require a change order. They include (Gould and Joyce, 2002; Schaufelberger and Holm, 2002):

1. Correction of design errors and discrepancies
2. Request from owner for changes and modifications to the project
3. Discovery of changed site or subsurface conditions
4. Delays affecting the completion date
5. Changes in regulatory requirements

Change orders are a major cause of disputes on construction projects. There may be disagreements about when a change order is necessary and about the amount of compensation a contractor is entitled to. Change orders require a negotiation between the owner and the contractor to establish what contract modifications are required. Often, it is difficult to establish the effects of a change on project costs and the project schedule. Contractors often do not like change orders because they are disruptive to their workflow, and the reallocation of resources to the work that the change order will entail may have undesirable schedule implications. Owners are always concerned that change orders will cause the cost of a project to spiral out of control.

A major cause of change orders is when changed conditions are encountered in subsurface work. Often, soil or rock conditions differ from those anticipated in the design. If the design documents do not provide an indication of the conditions encountered during construction, the contractor will seek a change order to cover the unanticipated costs.

When the need for a change order is determined, the designer produces documentation that describes the scope of the change order and documentation such as necessary plans, sketches, and specifications. The contractor examines this proposal request and negotiates with the owner on the cost to perform the additional work. When the contractor and owner agree on the cost of the change, a formal change order is issued that modifies the original construction contract (Gould and Joyce, 2002). When the owner and contractor cannot agree on the cost, the contractor is still obligated to perform the work, but the cost of the change order will become a claim that will be settled after the project is complete.

In some instances, change orders can be highly profitable for contractors. There is no competition in the negotiation with the owner, as during the initial bidding process; so the contractor can charge higher prices for the work than when competing with other contractors. Experienced contractors sometimes bid low on projects they know have poor designs in the hope of making money on the many change orders that will be required (Halpin, 2006).

Time Extensions

There are various reasons for time extensions. They include:

1. Delays resulting from unexpectedly severe weather
2. Delays caused by the owner such as interference with the work site
3. Delays caused by slow responses from the designer in processing submittals and requests for information from the field
4. Delays caused by external events beyond the contractor's control, such as strikes and environmental protests

A change order for a time extension may include only additional time to complete the project or it may also include additional funds to cover the contractor's costs if it can be proven that actions of the owner or designer caused additional expenditures for the contractor. The contractor can receive more funds if he or she documents the actions of the owner and designer that delayed the project during the change order negotiations.

The Ethics of Competitive Bidding

Although competitive bidding is usually done to insure fairness in the selection of contractors, unethical practices do occur. Bidders can illegally collude in the preparation of bids. Collusion is defined as an illegal, secret agreement between two or more parties for a fraudulent or wrongful purpose (AASHTO Subcommittee on Construction, 2003). A common form of collusion in the construction industry is the practice of bid rigging. Bid rigging occurs when competitors agree in advance who will win a contract. Contractors appear to be competing with each other when, in fact, they agree to cooperate on the winning bid to increase job profit (Peters, 2006).

Bid rigging typically takes four different forms (AASHTO Subcommittee on Construction, 2003):

- **Bid suppression.** Some contractors who normally would have bid refrain from bidding or withdraw a submitted bid.
- **Complementary bidding.** There is a predetermined winning contractor and the cooperating contractors submit bids that are too high.
- **Bid rotation.** The colluding contractors take turns being the winning low bidder.
- **Subcontracting.** The winning contractor gives large subcontracts to other collusive bidders in exchange for their not submitting a winning bid.

Collusive bidding practices are both illegal and unethical. Punishments for bid rigging include jail time and monetary penalties. Companies can be fined and disbarred from participating in government contracts. Government agencies actively seek to prosecute those involved in this type of fraud. For example, the Office of the Inspector General of the U.S. Department of Transportation has indicted 301 people and obtained 234 convictions in the period from 2000 to 2005. Penalties have totaled $153 million in fines and restitution and 734 years in jail and probation time. If found guilty, unethical conduct in preparing competitive bids can ruin a company and ruin a construction manager's life and should be avoided at all costs.

Summary

This chapter has illustrated several important points that everyone involved in construction must understand. They are:

1. There are important differences in the way designers and contractors are selected by the owner. Designers are selected by merit. Contractors can be selected based on merit, but in publicly funded work, they are often selected through the use of competitive bidding.
2. The competitive bidding process starts with the bid advertisement and completes when the low bidder receives the notice to proceed.
3. The bid package, made up of the plans, specifications, general conditions, special conditions, and proposal form, defines the project requirements, including both what is to be constructed and the legal relationship between the owner and contractor.
4. Many construction owners have standard specifications and standard general conditions that they use repeatedly on their projects.
5. Contract changes during construction are made through the use of change orders. The change order process is the source of many disputes.
6. Unethical and illegal bidding practices such as collusion and bid rigging do occur. These activities carry heavy criminal and civil penalties. All contractors must avoid them.

Key Terms

Addenda
Bid bond
Bid opening
Bid package
Change orders
Claims
Contract
General conditions
Liquidated damages
Notice to proceed
Plans
Prequalify
Progress payments
Proposal form
Retainage
Special conditions
Specifications
Time extension

Review Questions

1. What documents describe the design of a construction project? How do they differ?
2. What is the difference between the general and special conditions?
3. Where are the method and frequency of progress payments defined in the construction contract documents?
4. In what document is the project duration specified?
5. Why do you think large construction organizations prefer to use standard specifications?
6. Are designers selected through competitive bidding? Typically, what is the basis for their selection?
7. What is bid rigging? Why is it unethical?

Management Pro

You have been asked to write a specification for rough carpentry for a wooden truss that is to be constructed. The owner is sensitive to environmental issues. The owner has asked that you allow the contractor to use recovered and salvaged timber on the project. Download the SpecIntact software (See the Web link given in the text.) and use the rough carpentry specification as a basis for your specification. Note how the software allows you to tailor the standard specification to your specific project.

References

AASHTO Subcommittee on Construction. 2003. Collusion in department of transportation contracts. Available from http://cms.transportation.org/sites/construction/docs/2003%20Collusion%20Presentation%20SOC.ppt (accessed January 15, 2007).

Chua, D.K.H. and D. Li. 2000. Key factors in bid reasoning model. *Journal of Construction Engineering and Management* 126 (5): 349–357.

Dyer, Douglas and John H. Kagel. 1996. Bidding in common value auctions: How the commercial construction industry corrects for the winner's curse. *Management Science* 42 (10): 1463–1475.

Gould, Frederick E. and Nancy E. Joyce. 2002. *Construction Project Management*. Upper Saddle River, NJ: Prentice-Hall.

Halpin, Daniel W. 2006. *Construction Management*, 3rd ed. Hoboken, NJ: John Wiley.

Hana, Awad S., Richard Camlic, Pehr A. Peterson, and Erik V. Nordheim. 2002. Quantitative definition of projects impacted by change orders. *Journal of Construction Engineering and Management* 128 (1): 57–64.

Jackson, Babara J. 2004. *Construction Management Jump Start*. Alameda, CA: SYBEX.

Jervis, Bruce M. and Paul Levin. 1988. *Construction Law: Principles and Practice*. New York: McGraw-Hill.

Kelleher, Thomas J., ed. 2005. *Smith, Currie & Hancock's Common Sense Construction Law*, 3rd ed. Hoboken, NJ: John Wiley.

Peters, Mark. 2006. Highway fraud awareness. Available from www.transportation.org/sites/construction/docs/Peters%20-%20Highway%20Fraud%20Awareness.pdf (accessed January 15, 2007).

Schaufelberger, John E. and Len Holm. 2002. *Management of Construction Projects*. Upper Saddle River, NJ: Prentice Hall.

SpecsIntact. 2007. SpecsIntact: Future solutions now. Available from http://specsintact.ksc.nasa.gov/PDF/SIOverviewBrochure.pdf (accessed March 1, 2007).

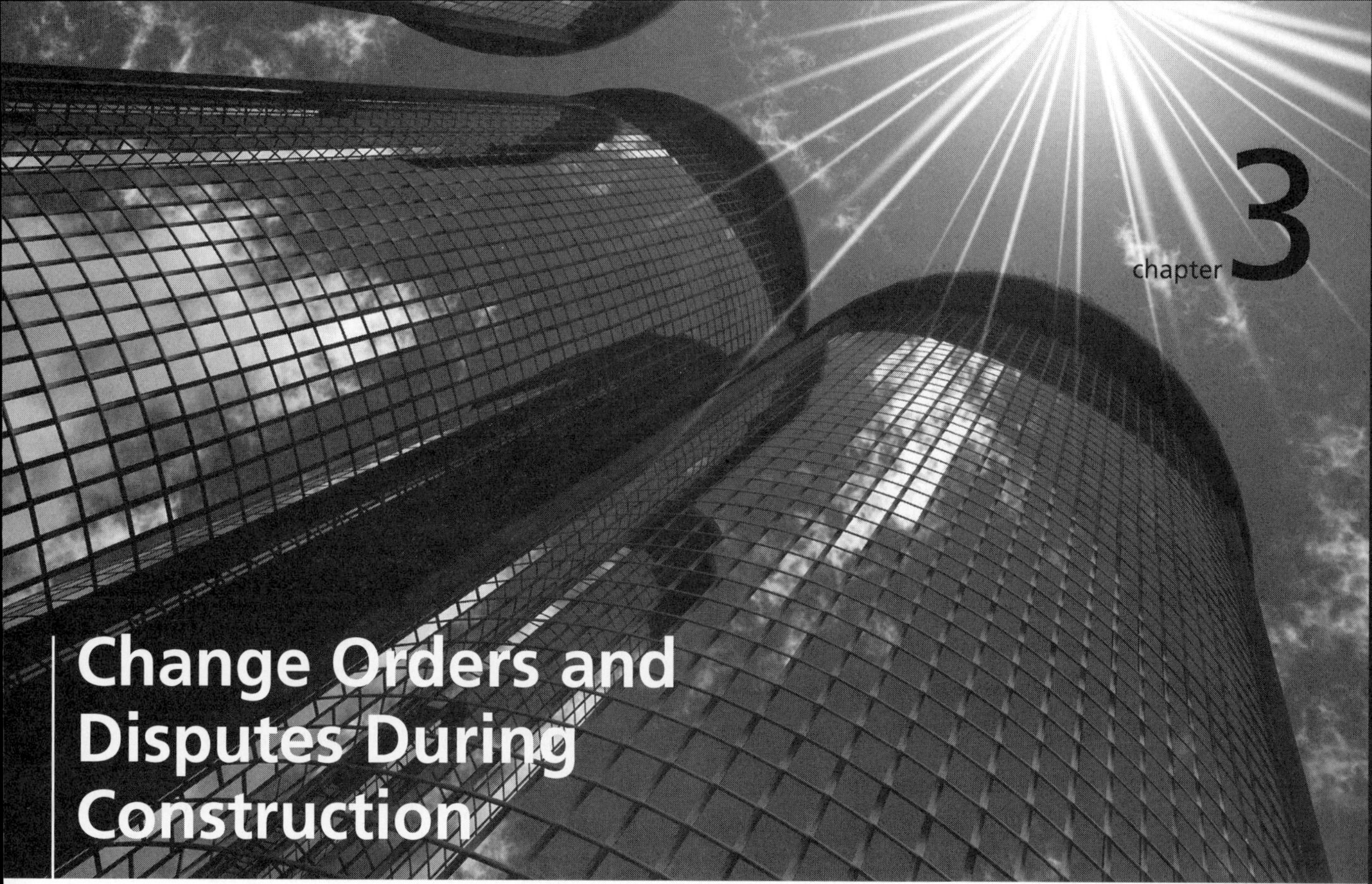

chapter 3

Change Orders and Disputes During Construction

Chapter Outline

Introduction

When you open the newspaper every morning, it is obvious that the world is filled with disputes and misunderstandings. In construction, we ask a contractor to build a complex facility from plans and specifications that may be difficult to interpret. Naturally, many disputes arise in construction. In Chapter 2, we noted that the construction contract, usually in the general conditions, provides rules for handling changes in the project during construction. In this chapter, we will discuss change orders in more detail. Often, contractors and owners are unable to reconcile their differences using change orders. We will examine how projects are completed when no agreement can be reached, and we will discuss construction claims, arbitration/mediation, and litigation.

Change Orders

In Chapter 2, we discussed that a change order clause is included in the construction contract documents. In this section, we will delve more deeply into the causes of change orders and their effect on a project. A change order is a written agreement between the owner and the contractor that changes the construction contract. A change order can add to, delete from, or otherwise alter the work described in the original contract documents. There are many reasons for initiating a change order. As we have already stated, problems often occur during construction that require a change order. Typical causes of change orders are (Joint Legislative Committee on Performance Evaluation and Expenditure Review 2002):

- **Unforeseen conditions.** Site conditions may differ from those expected. If so, a change order can be requested by the contractor or the designer.
- **Errors and omissions.** When errors and omissions are detected in the design, either the contractor or the designer can request a change order. Errors requiring a change order involve errors in the plans and/or specifications. Omissions are those required items or project elements that have been inadvertently omitted from the design documents.
- **Change in project design.** The owner has requested a change in the design. The owner may realize during construction that his or her requirements for the facility have changed, and that a modification in the design is necessary.

Figure 3.1 shows a contract clause related to change orders from the Federal Acquisition Regulations (FAR 53.243-4). This contract clause states the type of changes that the owner, in this case a federal agency, may request. Notice that it is the contractor's responsibility to notify the owner in writing that a modification to the contract price is required even though the owner may initiate the change. The contractor must file a request for the change order within 30 days of being informed of the change by the owner.

Example

The Federal Acquisition Regulations contain the text of all contract clauses that would be included in the general conditions of a federally funded construction project. The FAR Web site provides online access to all the contract clauses related to construction. The site is an excellent way to explore how construction contracts are written. A link to the regulations can be accessed at the federal government's Acquisition Central Web site at www.arnet.gov.

CHANGES (AUG 1987)

(a) The Contracting Officer may, at any time, without notice to the sureties, if any, by written order designated or indicated to be a change order, make changes in the work within the general scope of the contract, including changes—

(1) In the specifications (including drawings and designs);

(2) In the method or manner of performance of the work;

(3) In the Government-furnished facilities, equipment, materials, services, or site; or

(4) Directing acceleration in the performance of the work.

(b) Any other written or oral order (which, as used in this paragraph (b), includes direction, instruction, interpretation, or determination) from the Contracting Officer that causes a change shall be treated as a change order under this clause; Provided, that the Contractor gives the Contracting Officer written notice stating—

(1) The date, circumstances, and source of the order; and

(2) That the Contractor regards the order as a change order.

(c) Except as provided in this clause, no order, statement, or conduct of the Contracting Officer shall be treated as a change under this clause or entitle the Contractor to an equitable adjustment.

(d) If any change under this clause causes an increase or decrease in the Contractor's cost of, or the time required for, the performance of any part of the work under this contract, whether or not changed by any such order, the Contracting Officer shall make an equitable adjustment and modify the contract in writing. However, except for an adjustment based on defective specifications, no adjustment for any change under paragraph (b) of this clause shall be made for any costs incurred more than 20 days before the Contractor gives written notice as required. In the case of defective specifications for which the Government is responsible, the equitable adjustment shall include any increased cost reasonably incurred by the Contractor in attempting to comply with the defective specifications.

(e) The Contractor must assert its right to an adjustment under this clause within 30 days after (1) receipt of a written change order under paragraph (a) of this clause or (2) the furnishing of a written notice under paragraph (b) of this clause, by submitting to the Contracting Officer a written statement describing the general nature and amount of the proposal, unless this period is extended by the Government. The statement of proposal for adjustment may be included in the notice under paragraph (b) of this clause.

(f) No proposal by the Contractor for an equitable adjustment shall be allowed if asserted after final payment under this contract.

Figure 3.1 Change order general conditions from Federal Acquisition Regulations

Changed Conditions

A characteristic of many types of construction is that work is done underground where the exact nature of the underlying soil and rock may be unknown until construction actually starts. Examples include building foundations and subway tunnels. Changed subsurface conditions are the cause of many legal disputes. Typically, a contractor may request a change order for a **changed condition** if the conditions encountered differ materially from what is indicated in the design documents or if the conditions encountered are of an unusual nature that differ materially from conditions that are ordinarily encountered (Levin, 1998). Here are two examples of changed conditions:

- A contractor finds that the rock encountered during the construction of a subway tunnel is much different than the material described in the design documents. The contractor requests a change order for the cost of working in a more difficult material. This is an example of a changed condition where the conditions differ materially from the design documents.
- A slightly different situation is when the conditions encountered differ materially from conditions that might normally be expected. One example would be the discovery that polluted groundwater has been identified at the project site. Further investigation reveals that the source of the pollution is from an off-site chemical spill. The contractor can request a change order for the additional expenses to deal with the hazardous materials.
- Figure 3.2 shows the contract clause from the Federal Acquisition Regulations (FAR 52.243-5) for changed conditions. This contract clause shows that the contractor has a responsibility to immediately inform the owner when changed conditions are encountered.

Design Errors and Omissions

A common cause of change orders is errors and omissions in the plans and specifications. Given the complexity of projects constructed, plans and specifications are bound to contain errors. When such an error occurs, a change order must be prepared to fix the error. Additional costs to the contractor can include the cost of removing flawed work and the costs associated with applying different construction techniques to the redesigned work.

An example of the need for a change order could be the following situation. During construction, a contractor notices that the concrete specified for the project is not strong enough and that cracks are appearing in the building. The contractor realizes that there is a design error and requests the owner for redesign of the concrete, and a change order for removal of the substandard concrete and its replacement with higher strength concrete. It should be noted that a contractor has a legal duty to inform the owner immediately when finding issues with a design that may affect the safety of the public.

CHANGES AND CHANGED CONDITIONS (APR 1984)

(a) The Contracting Officer may, in writing, order changes in the drawings and specifications within the general scope of the contract.

(b) The Contractor shall promptly notify the Contracting Officer, in writing, of subsurface or latent physical conditions differing materially from those indicated in this contract or unknown unusual physical conditions at the site before proceeding with the work.

(c) If changes under paragraph (a) or conditions under paragraph (b) increase or decrease the cost of, or time required for performing the work, the Contracting Officer shall make an equitable adjustment (see paragraph (d)) upon submittal of a "proposal for adjustment" (hereafter referred to as proposal) by the Contractor before final payment under the contract.

(d) The Contracting Officer shall not make an equitable adjustment under paragraph (b) unless—

(1) The Contractor has submitted and the Contracting Officer has received the required written notice; or

(2) The Contracting Officer waives the requirement for the written notice.

(e) Failure to agree to any adjustment shall be a dispute under the Disputes clause.

Figure 3.2 Definition of a changed condition

Change in Project Design

As stated in the change order contract clause shown in Figure 3.1, the owner may request changes in the drawings and specifications that are within the general scope of the original project. This often occurs as the owner realizes that some elements of the design are not what he or she had envisioned for the constructed facility. The author worked on a traffic signal and intersection reconstruction project where input from the public caused the owner to require a change order. A downtown improvement society suggested to the owner, a large city, that painting the traffic signal cabinets in a bronze color would be more aesthetically pleasing than white. The owner agreed and a change order was negotiated with the contractor to repaint the cabinets.

Negotiating Change Orders

Most construction contracts provide three basic means for pricing change orders, with choice being left to the owner. They include:

- **An agreed fixed amount.** A fixed price for the change order work is negotiated between the owner and the contractor.
- **Reimbursement for time and materials expended.** This is often called force account work. It requires the contractor to keep records that document the labor and materials expended for the change order work.
- **A price that is developed by the contractor after the change order work has commenced.** This is used when the change order work is a new type of construction and the contractor may have difficulty pricing the work in advance.

The price for a change order is developed through negotiation between the contractor and the owner. The contractor researches the issues involved in the change order and proposes a price to the owner. They negotiate and hopefully come to an agreement on an equitable price. The negotiation of a change order can be simple or it can be very complex. It becomes complex when the change order involves costs to the contractor above the cost of material and labor to complete the additional work. In particular, a change order may have impacts on other aspects of the project. The wording of change order clauses in the general conditions typically allow for the contractor to be compensated for impacts to unchanged work.

The factors that must be considered in determining a change order price include direct costs, which are the price of labor, equipment, and materials, and impact costs that encompass the cost impacts implementing the change order will have on the contractors other activities. Potential impact costs include:

- Impacts on other unchanged activities
- Loss of productivity due to reassignment of crews to new tasks, and loss of work rhythm
- Escalation of material prices
- Additional costs and disruption of expediting work activities

Contractor's Duty to Perform Change Order Work

A construction contractor cannot refuse to perform change order work even if no agreement on the pricing or scope of the change order can be reached. Construction contracts typically state that the contractor has a duty to perform all work requested by the owner. Exceptions would be work that is clearly beyond the scope of the original contract and work where no clear instructions have been received by the contractor (Levin, 1998). When a change order cannot be negotiated, the contractor has various alternatives that can be pursued both during the project and after it is completed.

Issues Related to Time and Delay

Construction projects often become delayed. Sometimes, a special type of change order, a time extension, is given that increases the project duration because of unforeseeable conditions, such as a natural disaster, which have caused delays beyond the contractor's control. These time extensions are important because construction contractors must pay a monetary charge called liquidated damages when they are late in finishing a project.

Liquidated Damages

As we discussed in Chapter 2, the length of the construction project is specified in the special conditions. The failure to complete the project within the time specified can have significant monetary implications for the contractor. Construction contracts include a monetary charge for late completion called liquidated damages (see Chapter 2). These charges are typically assessed on a daily basis. Figure 3.3 shows the liquidated clause from the Federal Acquisition Regulation (FAR 52.211-12).

LIQUIDATED DAMAGES—CONSTRUCTION (SEPT 2000)

(a) If the Contractor fails to complete the work within the time specified in the contract, the Contractor shall pay liquidated damages to the Government in the amount of ___________ [Contracting Officer insert amount] for each calendar day of delay until the work is completed or accepted.

(b) If the Government terminates the Contractor's right to proceed, liquidated damages will continue to accrue until the work is completed. These liquidated damages are in addition to excess costs of repurchase under the Termination clause.

Figure 3.3 FAR definition of liquidated damages

A charge for liquidated damages must reflect the cost to the owner of the loss of the use of the completed facility. The liquidated damage amount cannot be a penalty. Penalty clauses, when they are used, must include a bonus to the contractor for early completion. Owners usually include a penalty clause (sometimes called an incentive/disincentive clause) in the contract for projects that must be completed rapidly. A good example is a bridge replacement project on an important, heavily used highway.

Example

The New York State Thruway Authority used a penalty clause for the replacement of a damaged bridge on I-87 in Yonkers, NY. The contract specified a $5,000 bonus per day for early completion and a $5,000 dollar penalty per day for late completion, with a $50,000 maximum for either scenario. The contract finished 8 days early and the contractor was awarded a $40,000 bonus (Bai and Burkett William, 2006).

Time Extensions

We have already discussed the need for time extensions, which are change orders to deal with changes to the project's completion date. Figure 3.4 shows a time extension clause from section 52.211-13 of the Federal Acquisition Regulation.

There are two types of delays that occur in a construction project: excusable and nonexcusable. If the cause of the delay was something beyond the control of the

Time extensions for contract changes will depend upon the extent, if any, by which the changes cause delay in the completion of the various elements of construction. The change order granting the time extension may provide that the contract completion date will be extended only for those specific elements related to the changed work and that the remaining contract completion dates for all other portions of the work will not be altered. The change order may also provide an equitable readjustment of liquidated damages under the new completion schedule.

Figure 3.4 Time extensions

contractor, such as a natural disaster, then it is an excusable delay. In an excusable delay, the contractor is entitled to a time extension but no additional compensation. A nonexcusable delay is a delay for which the owner is responsible and the contractor may seek additional compensation as well as a time extension (Kelleher, 2005). Of course, delays that are the contractor's responsibility are not eligible for a time extension and the contractor will be subject to liquidated damages. It is interesting to note from Figure 3.4 that government regulations allow for the completion dates of delayed parts of the project to be extended while retaining the original completion date for project elements that are unaffected by delays.

Resolving Construction Disputes

If a change order cannot be successfully negotiated, various alternatives are open to the contractor and owner. Traditionally, a meeting was held at the end of projects to allow the contractor to present formal, documented claims to the owner. The owner and the contractor negotiate and the owner may pay the contractor additional funds. If the contractor is not satisfied with the settlement offered, he or she could decide to litigate the claim in court. This can be an expensive and time-consuming process involving extensive legal costs. A major drawback is that the outcome of a trial can be difficult to predict. As a result, various alternative dispute resolution techniques have evolved and are widely used in the construction industry. These include dispute resolution boards, arbitration, and mediation. It should also be noted that it is more difficult to pursue litigation against federal agencies, which often require a series of claims hearings before a litigant can appeal to a federal court. Figure 3.5 shows the contract clause from the Federal Acquisition Regulations for disputes. Examination of this clause indicates that the owner, in this case the Federal Government, encourages the contractor to use alternative dispute resolution methods, which will be discussed in the following text.

Dispute Resolution Boards and On-Site Neutrals

Dispute resolution boards (DRBs) are one method of resolving disagreements that occur during construction. A DRB consists of three neutral persons appointed jointly by the owner and contractor, who become part of the project team. Expenses for the establishment of the DRB are shared by all the parties involved in the construction project. The DRB is usually described in the construction contract documents and is organized at the beginning before there are any disputes. DRB members attend periodic meetings, review project documents, and assist in the resolution of problems as they occur during a construction project. Disputes that arise during a project are presented to the DRB in a hearing. The DRB's decision is typically represented as a recommendation or a nonbinding decision (American Arbitration Association, 2004). For smaller projects, a single neutral party may be used rather than a board of three, to minimize expenses.

The contractor and the owner may also use an on-site neutral to resolve disputes. This is a person hired with extensive knowledge and involvement with the project who acts

DISPUTES (DEC 1998)

(a) This contract is subject to the Contract Disputes Act of 1978, as amended (41 U.S.C. 601-613).

(b) Except as provided in the Act, all disputes arising under or relating to this contract shall be resolved under this clause.

(c) "Claim," as used in this clause, means a written demand or written assertion by one of the contracting parties seeking, as a matter of right, the payment of money in a sum certain, the adjustment or interpretation of contract terms, or other relief arising under or relating to this contract. A claim arising under a contract, unlike a claim relating to that contract, is a claim that can be resolved under a contract clause that provides for the relief sought by the claimant. However, a written demand or written assertion by the Contractor seeking the payment of money exceeding $100,000 is not a claim under the Act until certified as required by paragraph (d)(2) of this clause. A voucher, invoice, or other routine request for payment that is not in dispute when submitted is not a claim under the Act. The submission may be converted to a claim under the Act, by complying with the submission and certification requirements of this clause, if it is disputed either as to liability or amount or is not acted upon in a reasonable time.

(d) (1) A claim by the Contractor shall be made in writing and, unless otherwise stated in this contract, submitted within 6 years after accrual of the claim to the Contracting Officer for a written decision. A claim by the Government against the Contractor shall be subject to a written decision by the Contracting Officer.

(2) (i) The Contractor shall provide the certification specified in paragraph (d)(2)(iii) of this clause when submitting any claim exceeding $100,000.

(ii) The certification requirement does not apply to issues in controversy that have not been submitted as all or part of a claim.

(iii) The certification shall state as follows: "I certify that the claim is made in good faith; that the supporting data are accurate and complete to the best of my knowledge and belief; that the amount requested accurately reflects the contract adjustment for which the Contractor believes the Government is liable; and that I am duly authorized to certify the claim on behalf of the Contractor."

(3) The certification may be executed by any person duly authorized to bind the Contractor with respect to the claim.

(e) For Contractor claims of $100,000 or less, the Contracting Officer must, if requested in writing by the Contractor, render a decision within 60 days of the request. For Contractor-certified claims over $100,000, the Contracting Officer must, within 60 days, decide the claim or notify the Contractor of the date by which the decision will be made.

(f) The Contracting Officer's decision shall be final unless the Contractor appeals or files a suit as provided in the Act.

(g) If the claim by the Contractor is submitted to the Contracting Officer or a claim by the Government is presented to the Contractor, the parties, by mutual consent, may agree to use alternative dispute resolution (ADR). If the Contractor refuses an offer for ADR, the Contractor shall inform the Contracting Officer, in writing, of the Contractor's specific reasons for rejecting the offer.

(h) The Government shall pay interest on the amount found due and unpaid from (1) the date that the Contracting Officer receives the claim (certified, if required); or (2) the date that payment otherwise would be due, if that date is later, until the date of payment. With regard to claims having defective certifications, as defined in FAR 33.201, interest shall be paid from the date that the Contracting Officer initially receives the claim. Simple interest on claims shall be paid at the rate, fixed by the Secretary of the Treasury as provided in the Act, which is applicable to the period during which the Contracting Officer receives the claim and then at the rate applicable for each 6-month period as fixed by the Treasury Secretary during the pendency of the claim.

(i) The Contractor shall proceed diligently with performance of this contract, pending final resolution of any request for relief, claim, appeal, or action arising under the contract, and comply with any decision of the Contracting Officer.

Figure 3.5 Definition of disputes and claims

as a mediator or helps find ways to deal with difficult construction problems. Typically, an on-site neutral would have much more extensive exposure to a project, possibly on a day-to-day basis, than a single party neutral who would be called in only when a dispute needed to be resolved (AAA, 2006).

Example

Alternative Dispute Resolutions are the techniques that are available to settle construction disputes without resorting to a court trial. They have become widely used in the construction industry because they are usually less costly and less time consuming than a court trial. Supplemental material about dispute resolution methods is available at the Web site of the American Arbitration Association at www.adr.org

Arbitration and Mediation

Often it is not possible to settle disputes during construction. **Arbitration** and **mediation** are often used as tools to help settle disputes and to resolve outstanding change order claims. Arbitration can be defined as the hearing and determination of a dispute by an impartial referee agreed to by the owner and the construction contractor. Mediation is defined as a negotiation to resolve differences that is conducted by some impartial party (Princeton University, 2006).

In the construction industry, a single arbitrator is sometimes used, or a panel of three construction experts is used. The arbitrators are neutral parties with no connections with any of the parties in the project. The American Arbitration Association (AAA) provides impartial dispute resolution. The AAA maintains lists of qualified arbitrators and can assist in facilitating arbitration hearings. When arbitration is to be used in a project, language will be inserted in the contract documents defining the procedures. Typically, the arbitration decision is made binding on both parties.

Arbitration is often used in construction because:

- It can be faster and less costly than judicial proceedings.
- The dispute is presented to a panel of industry experts rather than a jury of lay people.
- It is private and not subject to public disclosure like many judicial proceedings.

A formal hearing is typically held where the arbitrators hear presentations, receive evidence, and ask questions about the claim. An arbitration hearing is less formal than a court trial, although attorneys may present the case for each side. The arbitrators then decide on a solution. Depending on the construction contract, the arbitrator's decision may be binding on the owner and the contractor or just advisory.

Mediation is another possibility to resolve outstanding construction claims. In the construction industry, one or more mediators may be used to impartially guide the owner and the contractor to reach a solution. The mediator facilitates negotiations and consults with the parties in an effort to find a solution to the claim that is agreeable to all parties. The mediator does not impose a settlement, and participation is normally an informal and voluntary process (American Arbitration Association, 2004). Mediation is sometimes used as a method of avoiding arbitration or litigation.

Litigation

Litigation can be defined as a contest authorized by law, in a court of justice, for the purpose of enforcing a right. In other words, litigation is a court trial where contractors can sue owners (or vice versa) to receive additional funds or legal redress they believe they are entitled to. A contractor may pursue litigation against an owner, for example, if he or she believes that payment made for change order work is not enough. Alternatively, an owner could sue a contractor because he or she believes the work is of a substandard quality.

Litigation is typically the solution of a last resort to disputes that occur during construction. The wording of the general conditions varies widely among construction projects, but most now include language that encourages alternative dispute resolution before litigation. Some construct contracts are written that require disputes be submitted for arbitration rather than litigation. Figure 3.5 shows that for government contracts, a contractor must explain in writing why he or she refuses to try alternative dispute resolution.

One study (Cruz, n.d.) has provided a comparison of the procedures, time, and cost required to pursue arbitration versus litigation on two similar projects. Major differences were noted in time and legal expense:

- An arbitration of $25 million in claims against a contractor in New Jersey was settled in a little over 2 years. Total legal, arbitration, and American Arbitration Association fees were around $1,125,000.
- The second case related to a contractor's work at a plant in New York. The case involved $200 million in claims and was litigated in court. The case took 6 years to settle and the legal fees totaled more than $7 million (American Arbitration Association, 2004).

As this example indicates, litigation can potentially be longer and costlier than arbitration. A wise contractor will carefully consider all possible avenues to dispute resolution before pursuing litigation.

Case Study: An Example of a Construction Claim that Resulted in Complex Litigation

The following is an article about a claim for the construction of a subway tunnel. Students should note that the litigation has taken 11 years and is still continuing. The litigation is complex with claims and cross-claims between the owner and the contractor. The claim involves large sums of money and illustrates the potential consequences to a contractor of losing in a trial. This article, titled "California Jury Says Tutor-Saliba Breached Contract on Tunnel," was published in *ENR* magazine on December 25, 2006. Tony Illia wrote the article.

> The 11-year legal battle between the Los Angeles County Metropolitan Transit Authority (MTA) and its contractor, Tutor-Saliba-Perini, over construction of the Red Line subway, took a new turn on Dec. 18. A state Superior Court jury found the joint venture guilty of submitting false claims and breach of contract. It awarded MTA $446,604 in damages.
>
> The new decision rules on 8,800 ft of disputed hand-railing inside adjoining tunnels at the Wilshire-Normandie subway station. MTA issued and paid for a change order related to the railing. But Tutor-Saliba-Perini claimed it as added work due to ambiguous bid documents and architectural plans.
>
> "MTA has made this fight a matter of principle and not money," says Ronald Helmuth, a Pasadena-based attorney representing the agency. "It's appropriate to bring this type of wrongful conduct to light." Calls to Ronald N. Tutor, president of Tutor-Saliba Corp., Sylmar, Calif., were not returned by *ENR* press time.
>
> In 1995, Tutor-Saliba-Perini filed suit to recover $3 million in claimed work on the Wilshire-Normandie station. Its $79-million contract had increased by $20 million based on owner-approved change orders. MTA, in response, filed a cross complaint in 1999.
>
> In 2001, a Superior Court judge cut proceedings short and found in favor of MTA, sanctioning Tutor-Saliba-Perini for allegedly destroying and/or concealing vital documents. MTA was awarded $60 million. But the award was overturned on appeal in 2005 in favor of the joint venture. An appellate court ruled that the lower court judge abused his discretion and violated the contractor's due process rights.
>
> Yet the bitter courtroom fight is far from over. Superior Court Judge Carolyn Kuhl now will rule on up to $82,500 in penalties for Tutor-Saliba-Perini's violation of the state's unfair competition law. About $20 million in claims and counterclaims also will be decided [in] trials set to start next spring.

Communications and Partnering

One of the best ways to avoid disputes during and after a construction project is to provide open lines of communication between project participants to quickly solve problems and disputes before they lead to the need for costly arbitration or litigation.

The *Guide to the Project Management Body of Knowledge* (Project Management Institute, 2008) lists several processes for managing project communications. First, it is important to identify project stakeholders. In construction, the primary stakeholders are the owner, the designer, and the contractor, but many other parties such as government agencies and the general public may have stakes in a complex construction project.

Planning communication needs is the next step in managing project communications once the stakeholders have been identified. What information each stakeholder needs must be identified. Construction projects require good communications both internally in the stakeholders' organization and also externally with other stakeholders. For successful project communications, it is necessary to plan so that:

- All the necessary communications reach the correct recipient
- The information is sent in a timely manner
- Responsibilities are clearly defined in each stakeholder's organization concerning who is responsible for the sharing and flow of information

A solution that has emerged to enhance project communications and to reduce the number of construction disputes is to promote formal **partnering** between the parties of a construction project. The American Arbitration Association's Dispute Avoidance and Resolution Task Force (1996) has defined partnering as a voluntary organized process by which the parties to a construction contract perform as a team to achieve mutually beneficial goals. Partnering is a collaborative process and it promotes increased interaction between construction project participants. Formal partnering is a nonbinding process and does not involve any contract document changes. Instead, partnering is a commitment by the parties to cooperate with each other.

Implementation of successful partnering involves several phases. They include:

- **Phase 1.** Project goals and long-term plans for implementation are defined.
- **Phase 2.** Support of project participants including high-level management is enlisted. Training in partnering techniques is undertaken.
- **Phase 3.** Teams that are formed will work together to implement the project. Problem-solving processes are set up. A project charter is written at this point that defines project goals and the teams' commitment to work together.

- **Phase 4.** This phase occurs during construction. At this stage, the partners engage in daily collaborative problem solving. There are numerous team meetings, and difficult technical issues are resolved. Creative thinking by all team members is encouraged.
- **Phase 5.** Partners work together to close out the project and identify successes that can be incorporated in future construction projects.

Typically, a knowledgeable facilitator is employed to run partnering meetings, and to educate the participants about partnering techniques and practices. Potentially, partnering can allow the construction industry to deliver improved value by providing techniques to address construction problems at an early stage and by resolving the problems rapidly through improved cooperation between the contractor, owner, and designer.

Value Engineering

Value engineering is the analysis of a construction project to more clearly define its scope and to identify potential cost savings. It has also been defined as a systematic review of a project by a team of experts to improve performance, quality, and life-cycle costs (Wilson, 2005). Value engineering involves the use of creative thinking to provide alternative methods for constructing a project at a lower cost while retaining the functionality of the original design. It is used widely by government agencies. State transportation agencies are required to use it on federally funded projects greater than $25 million.

Example

Value engineering is the evaluation of a design to determine if it can be modified to produce the same functionality as the original design but at a lower cost. Extensive supplementary material about value engineering is provided by the Federal Highway Administration at www.fhwa.dot.gov/ve/.

Value engineering can be implemented in two ways. One method is for a team from the owner's organization to be formed to review the design before the project is advertised for bidding. This team makes changes to the project design and after the modifications are made, the project is advertised for bid.

Closely related to value engineering is the concept of constructability. Designers often may lack experience in construction techniques and may produce a design that is difficult for the contractor to construct. Designs that are difficult to construct will take more time and may be more costly. An example is in bridge construction where it is possible to design complex connections between structural members. A constructability analysis will suggest alternative methods of making connections that will be easier for the welders in the field to build.

The second method is to allow the construction contractor to suggest changes to the design after the project has been awarded. In this form of value engineering, the construction contract contains a provision that encourages the contractor to propose changes in the project that provide the same functionality as the original design, but which can be constructed for a lower cost, or which will improve the project value at no increase in cost (Federal Highway Administration, 2006). Changes of this nature are often referred to as a value engineering change proposal (VECP). It should be noted that this is another form of a change order, albeit a positive one.

The VECP process includes the following steps:

- The contractor submits a VECP with ideas to reduce project costs.
- The owner agency reviews the feasibility of the VECP and its effect on the project. The owner decides to accept or reject the proposal.
- If the VECP is rejected, the contractor proceeds with construction according to the original design.
- If accepted, the contractor and the owner agency share the identified cost savings.

If a 50/50 cost sharing is agreed on, then a contractor finding a $100,000 saving in the project would receive half of the saving ($50,000) as a payment, and the project would cost the owner $50,000 less.

This method gives the contractor a financial incentive to seek innovative and efficient construction methods. Allowing the contractor to participate in value engineering also reduces the adversarial relationships that may develop on a construction project and promotes collaboration between the owner and contractor. In general, both methods of value engineering can have positive effects on a construction project by providing for a detailed review and assessment of the design to seek improvements in the project.

Figure 3.6 shows the value engineering clause for construction cost sharing from the Federal Acquisition Regulation (FAR 52.249-2) that would be included as a clause in the general conditions of the construction contract documents. This contract clause would allow a contractor in a fixed-price contract to recoup 55% of any cost savings plus 20% of any collateral savings that are generated. The contract clause also states that the contractor must include value engineering clauses in contracts he or she has with subcontractors on subcontracts greater than $50,000.

VALUE ENGINEERING—CONSTRUCTION (FEB 2000)

(a) *General.* The Contractor is encouraged to develop, prepare, and submit value engineering change proposals (VECP's) voluntarily. The Contractor shall share in any instant contract savings realized from accepted VECP's, in accordance with paragraph (f) of this clause.

(b) *Definitions.* "Collateral costs," as used in this clause, means agency costs of operation, maintenance, logistic support, or Government-furnished property.

"Collateral savings," as used in this clause, means those measurable net reductions resulting from a VECP in the agency's overall projected collateral costs, exclusive of acquisition savings, whether or not the acquisition cost changes.

"Contractor's development and implementation costs," as used in this clause, means those costs the Contractor incurs on a VECP specifically in developing, testing, preparing, and submitting the VECP, as well as those costs the Contractor incurs to make the contractual changes required by Government acceptance of a VECP.

"Government costs," as used in this clause, means those agency costs that result directly from developing and implementing the VECP, such as any net increases in the cost of testing, operations, maintenance, and logistic support. The term does not include the normal administrative costs of processing the VECP.

"Instant contract savings," as used in this clause, means the estimated reduction in Contractor cost of performance resulting from acceptance of the VECP, minus allowable Contractor's development and implementation costs, including subcontractors' development and implementation costs (see paragraph (h) of this clause).

"Value engineering change proposal (VECP)" means a proposal that—

(1) Requires a change to this, the instant contract, to implement; and

(2) Results in reducing the contract price or estimated cost without impairing essential functions or characteristics; *provided*, that it does not involve a change—

(i) In deliverable end item quantities only; or

(ii) To the contract type only.

(c) *VECP preparation.* As a minimum, the Contractor shall include in each VECP the information described in paragraphs (c)(1) through (7) of this clause. If the proposed change is affected by contractually required configuration management or similar procedures, the instructions in those procedures relating to format, identification, and priority assignment shall govern VECP preparation. The VECP shall include the following:

(1) A description of the difference between the existing contract requirement and that proposed, the comparative advantages and disadvantages of each, a justification when an item's function or characteristics are being altered, and the effect of the change on the end item's performance.

(2) A list and analysis of the contract requirements that must be changed if the VECP is accepted, including any suggested specification revisions.

(3) A separate, detailed cost estimate for (i) the affected portions of the existing contract requirement and (ii) the VECP. The cost reduction associated with the VECP shall take into account the Contractor's allowable development and implementation costs, including any amount attributable to subcontracts under paragraph (h) of this clause.

(4) A description and estimate of costs the Government may incur in implementing the VECP, such as test and evaluation and operating and support costs.

(5) A prediction of any effects the proposed change would have on collateral costs to the agency.

(6) A statement of the time by which a contract modification accepting the VECP must be issued in order to achieve the maximum cost reduction, noting any effect on the contract completion time or delivery schedule.

Figure 3.6 FAR value engineering clause

(7) Identification of any previous submissions of the VECP, including the dates submitted, the agencies and contract numbers involved, and previous Government actions, if known.

(d) *Submission.* The Contractor shall submit VECP's to the Resident Engineer at the worksite, with a copy to the Contracting Officer.

(e) Government action.

(1) The Contracting Officer will notify the Contractor of the status of the VECP within 45 calendar days after the contracting office receives it. If additional time is required, the Contracting Officer will notify the Contractor within the 45-day period and provide the reason for the delay and the expected date of the decision. The Government will process VECP's expeditiously; however, it will not be liable for any delay in acting upon a VECP.

(2) If the VECP is not accepted, the Contracting Officer will notify the Contractor in writing, explaining the reasons for rejection. The Contractor may withdraw any VECP, in whole or in part, at any time before it is accepted by the Government. The Contracting Officer may require that the Contractor provide written notification before undertaking significant expenditures for VECP effort.

(3) Any VECP may be accepted, in whole or in part, by the Contracting Officer's award of a modification to this contract citing this clause. The Contracting Officer may accept the VECP, even though an agreement on price reduction has not been reached, by issuing the Contractor a notice to proceed with the change. Until a notice to proceed is issued or a contract modification applies a VECP to this contract, the Contractor shall perform in accordance with the existing contract. The decision to accept or reject all or part of any VECP is a unilateral decision made solely at the discretion of the Contracting Officer.

(f) Sharing—(1) *Rates.* The Government's share of savings is determined by subtracting Government costs from instant contract savings and multiplying the result by—

(i) 45 percent for fixed-price contracts; or

(ii) 75 percent for cost-reimbursement contracts.

(2) *Payment.* Payment of any share due the Contractor for use of a VECP on this contract shall be authorized by a modification to this contract to—

(i) Accept the VECP;

(ii) Reduce the contract price or estimated cost by the amount of instant contract savings; and

(iii) Provide the Contractor's share of savings by adding the amount calculated to the contract price or fee.

(g) *Collateral savings.* If a VECP is accepted, the Contracting Officer will increase the instant contract amount by 20 percent of any projected collateral savings determined to be realized in a typical year of use after subtracting any Government costs not previously offset. However, the Contractor's share of collateral savings will not exceed the contract's firm-fixed-price or estimated cost, at the time the VECP is accepted, or $100,000, whichever is greater. The Contracting Officer is the sole determiner of the amount of collateral savings.

(h) *Subcontracts.* The Contractor shall include an appropriate value engineering clause in any subcontract of $50,000 or more and may include one in subcontracts of lesser value. In computing any adjustment in this contract's price under paragraph (f) of this clause, the Contractor's allowable development and implementation costs shall include any subcontractor's allowable development and implementation costs clearly resulting from a VECP accepted by the Government under this contract, but shall exclude any value engineering incentive payments to a subcontractor. The Contractor may choose any arrangement for subcontractor value engineering incentive payments; *provided*, that these payments shall not reduce the Government's share of the savings resulting from the VECP.

Figure 3.6 *Continued*

(i) *Data.* The Contractor may restrict the Government's right to use any part of a VECP or the supporting data by marking the following legend on the affected parts:

> These data, furnished under the value engineering—Construction clause of contract __________, shall not be disclosed outside the Government or duplicated, used, or disclosed, in whole or in part, for any purpose other than to evaluate a value engineering change proposal submitted under the clause. This restriction does not limit the Government's right to use information contained in these data if it has been obtained or is otherwise available from the Contractor or from another source without limitations.

If a VECP is accepted, the Contractor hereby grants the Government unlimited rights in the VECP and supporting data, except that, with respect to data qualifying and submitted as limited rights technical data, the Government shall have the rights specified in the contract modification implementing the VECP and shall appropriately mark the data. (The terms "unlimited rights" and "limited rights" are defined in Part 27 of the Federal Acquisition Regulation.)

Figure 3.6 *Continued*

Summary

This chapter has provided a description of how changes and problems are dealt with during the construction process. It must be stressed that all the methods for change orders, time extensions, and methods of alternative dispute resolution are described in the general conditions of the bid package. Therefore, a contractor must have a thorough understanding of the applicable clauses both before deciding to pursue a project and during construction when disputes may arise.

The following important topics have been discussed:

1. Change orders are a formal modification to the original contract documents. They result from a negotiation between the owner and contractor. The need for change orders arises for a variety of reasons, including errors in the design documents, and owner-requested design changes.
2. If a change order cannot be negotiated, there are various methods of dispute resolution, including litigation, arbitration, and mediation. Litigation was the traditional method of resolving construction disputes but it may be costly and time consuming. Alternative methods of dispute resolution such as arbitration have become more widespread because construction disputes can potentially be settled more rapidly and with lower legal fees.
3. Methods that foster improved collaboration between the owner and the contractor can have a beneficial effect by reducing the adversarial relationships that sometime occur on construction projects. Partnering is a way of getting the various parties on a construction project to work together as a team. Value engineering allows contractors to be rewarded for innovating, by providing them with a share of project cost reductions for cost-saving modifications to the design.

Key Terms

- Arbitration
- Changed condition
- Dispute resolution boards (DRBs)
- Litigation
- Mediation
- Partnering
- Value engineering

Review Questions

1. What is the difference between arbitration and mediation?
2. Why does partnering help avoid construction disputes?
3. Construction claims can end in litigation. Why is litigation viewed as the solution of last resort?
4. When does a changed condition occur? What is the legal remedy?
5. Can a contractor refuse to do change order work? Explain.
6. How does value engineering allow cost sharing of savings during a project? What construction party identifies savings during construction?

MANAGEMENT PRO

You are the project manager for a large warehouse construction project. Design documents indicate that the underground soils are a soft loam. Therefore, you had anticipated it would be easy to excavate the foundation using conventional construction equipment. Instead, you find that the subsurface condition includes many large boulders that must be blasted using dynamite. The blasting operation is considerably more expensive than what you had bid for excavation. As project manager, what should you do to get increased compensation for the blasting?

References

American Arbitration Association. 2004. The construction industry's guide to dispute avoidance and resolution. Available from http://www.adr.org/si.asp?id=3839 (accessed April 15, 2007).

Bai, Yong and R. Burkett William. 2006. Rapid bridge replacement: Processes, techniques, and needs for improvements. *Journal of Construction Engineering and Management* 132 (11): 1139–1147.

Cruz, Jeffrey R. n.d. Arbitration vs. litigation: An unintentional experiment. American Arbitration Association. Available from http://www.adr.org/sp.asp?id=29193 (accessed March 15, 2007).

Dispute Avoidance and Resolution Task Force. 1996. Building success for the 21st century: A guide to partnering in the construction industry. American Arbitration Association. Available from http://www.adr.org/sp.asp?id=29169 (accessed April 15, 2007).

Federal Highway Administration. 2006. Construction program guide: Value engineering change proposals. Available from http://www.fhwa.dot.gov/construction/cqit/vecp.htm (accessed April 28, 2007).

Joint Legislative Committee on Performance Evaluation and Expenditure Review. 2002. The Bureau of Building's management of construction change orders. Report to the Mississippi Legislature #429. Available from http://www.peer.state.ms.us/reports/rpt429.pdf (accessed March 15, 2007).

Kelleher, Thomas J., ed. 2005. *Smith, Currie & Hancock's Common Sense Construction Law*, 3rd ed. Hoboken, NJ: John Wiley.

Levin, Paul. 1998. *Construction Contract Claims, Changes & Dispute Resolution*, 2nd ed. Reston, VA: ASCE Press.

Princeton University. 2006. WordNet 3.0. Available from http://wordnet.princeton.edu/perl/webwn?s=word-you-want (accessed March 15, 2007).

Project Management Institute. 2008. *Guide to the Project Management Body of Knowledge*, 4th ed. Newtown Square, PA: Project Management Institute.

Wilson, David C. 2005. *Value Engineering Applications in Transportation*. NCHRP Synthesis 352. Washington, DC: Transportation Research Board.

chapter 4

Construction Management Contracts: An Alternative to Design-Bid-Build

Chapter Outline

Introduction

Chapter 2 discussed the traditional approach to construction. In that approach, a project is designed, bid, and then constructed by the lowest bidder. Alternatively, the project is designed, and the owner negotiates with preferred contractors. These contract formats are still widely used. However, alternative construction contract types have also been developed on the basis of some of the perceived weaknesses of the traditional contract types. In this chapter, we will discuss how an owner can use a contract type that employs a construction management firm to provide advice.

One of the major problems with a traditional **design-bid-build** contract is that the design must be complete before the bidding can take place. For an owner with an urgent need to have a project completed, this process may take too long. **Fast-tracking** refers to the ability to compress the duration of a construction project by starting construction before the design is complete. Figure 4.1 shows a timeline for construction in traditional competitive bidding and the schedule compression that is possible by initiating construction before the design is complete. The total length of the design and construction phases is the same in the fast-track schedule, yet by dividing

Figure 4.1

Reduction in project duration due to fast-tracking

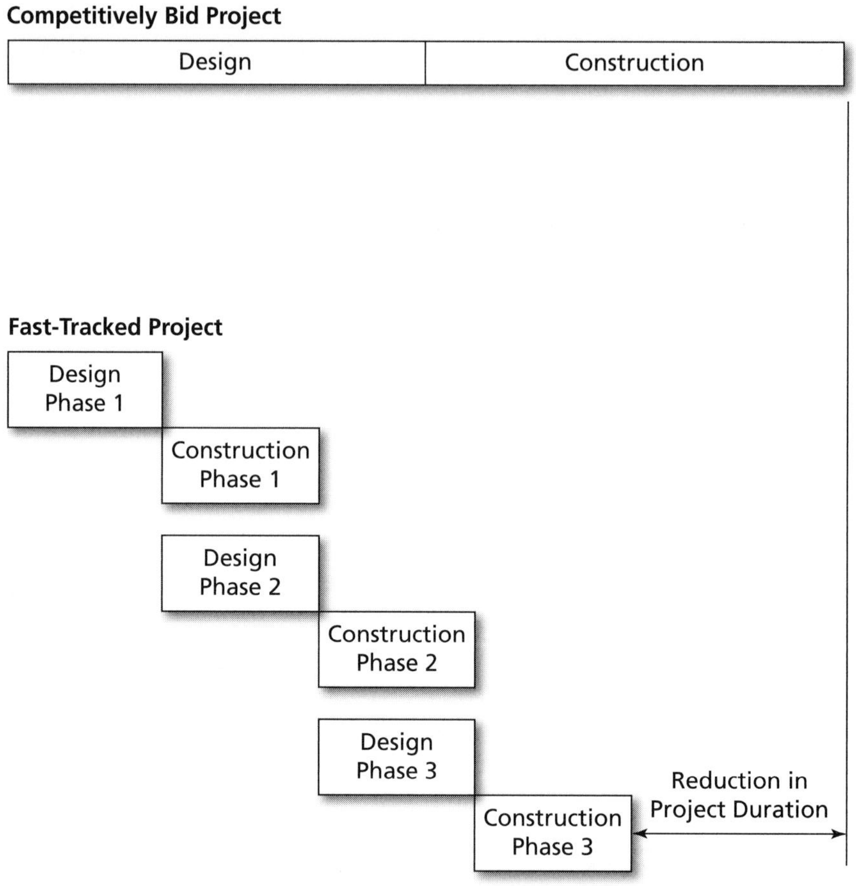

the project into three separate segments and starting construction after the first portion of the design is completed allows the schedule to be compressed significantly.

Another problem in traditional contracts is the lack of coordination between design and construction. The owner hires a designer who produces the plans and specifications without consultation with the contractor. It is generally accepted that enhanced communication between designers and experienced construction field personnel leads to fewer problems during construction. Finally, owners seek an improved arrangement, one in which the adversarial relationships that arise in the traditional contract format—friction between the designer and the owner, and the contractor and the owner—can be reduced.

A **construction manager** (CM) is a firm that provides professional services to an owner to assist in the planning, coordination, and construction of a facility. Many construction owners require complex facilities to be constructed, yet do not have their own full-time staff capable of overseeing the project. The concept of using a CM is to provide the owner with an advisor who can discuss complex technical issues with the designer and provide assistance in dealing with issues that arise during construction. The CM who is an advisor to the owner, is available from the initiation of a project to its completion.

Consider a facility like a hospital. The company that owns the hospital may have little internal knowledge about how to manage a construction project for a new addition. In this modern age, a hospital is a complex building to design and construct because it must accommodate high-tech medical and computer equipment. In this instance, a CM, contractual arrangement may be ideal for the hospital owner. The owner can hire a CM firm with experience in hospital projects to provide advice.

CMs make expert recommendations in the following areas:

- Value engineering and constructability analysis of the design
- Improved definition of the scope of the project
- Improved project scheduling and coordination
- Avoidance of delays, changes, and claims
- Design and construction quality (Construction Management Association of America, 2002)

A CM participates in a project from the inception of the design to project completion. To provide the expert advice that an owner requires, a CM will work with the owner and designer to define the scope and requirements of a project. In the hospital example, a CM will provide an intermediary that can consider the medical requirements of the hospital and insure that these requirements are incorporated in the engineering design. Unlike the traditional competitively bid project where the designer and contractor are essentially insulated from each other, the construction management contract format allows constructors to interact with designers to improve the project design.

Example

Additional information and documentation about the construction management format of construction is available on the Web page of the Construction Management Association of America at http://cmaanet.org/. This organization promotes professional construction management services as a professional discipline.

Construction Management Contract Formats

There are two types of construction management contracts. They are commonly referred to as **agency CM** and **CM at-risk (CMAR)**. Agency CM evolved in the 1970s. In this method, the CM acts solely as an advisor to the owner. Later, CMAR evolved. In CMAR, the CM takes on a role resembling that of a general contractor for the construction phase, while still providing advisory services to the owner during design. The at-risk method evolved because the CM using an agency CM is not contractually liable for the completion of the work. Although the agency CM is paid a fee by the owner, the latter has construction contracts only with the project contractors. Decisions made by the agency CM could have significant impacts on the project, yet the owner is liable. Poor decisions by an agency CM could result in increased costs for the owner (Halpin, 2006). Figure 4.2 provides a comparison of the contractual relationships found in CMAR and agency CM. The figure shows that the agency CM does not have a direct contractual link with the contractors who will be hired to perform the work. Yet, the agency CM is responsible for assisting the owner in bidding or negotiating the project. In contrast, the CMAR has a direct contractual link to the owner to build the project and another consulting contract for the services performed by the CM before construction.

Agency Construction Management

In agency CM, the CM provides expertise during the construction project (Construction Management Association of America, 2007), as listed here:

1. Pre-Design and Design
   - The CM assists the owner in selecting a design team.
   - The CM also assists the owner with preliminary cost estimates and budgets.
   - The CM conducts constructability and value engineering reviews of the design.
   - The CM assists the owner in conducting a competitive bid or negotiations with trade contractors. The CM often suggests contractors to the owner that it has had a good working relationship with.
   - The CM develops the bid package for each portion of the work. The CM works with the owner to decide how the project is to be broken down into contracts for the general contractor and subcontractors who will construct the project.

Figure 4.2

Differences in agency CM and CMAR contractual structures

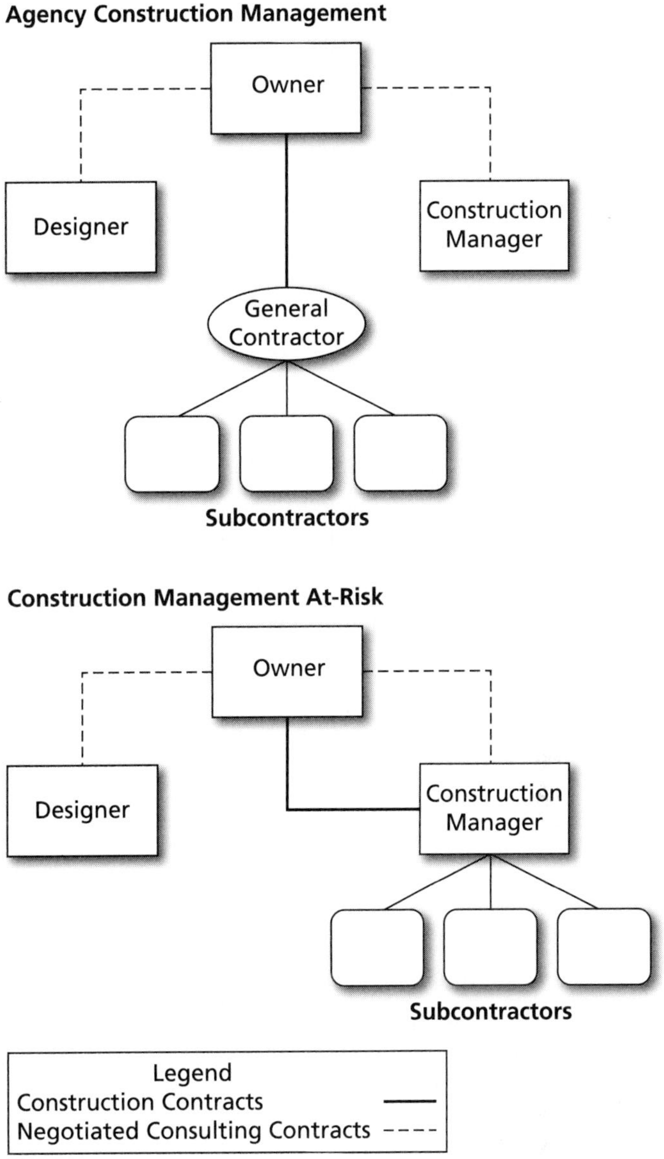

2. Construction
   - During construction, the CM provides construction inspection and surveillance.
   - The CM monitors construction for conformance with specifications and quality. The CM monitors the activities of each trade contractor.
   - A major function of the CM during construction is the management of project paperwork, correspondence, and meetings.
   - The CM reviews and processes change order requests.

Figure 4.3

Construction management activities throughout project life cycle

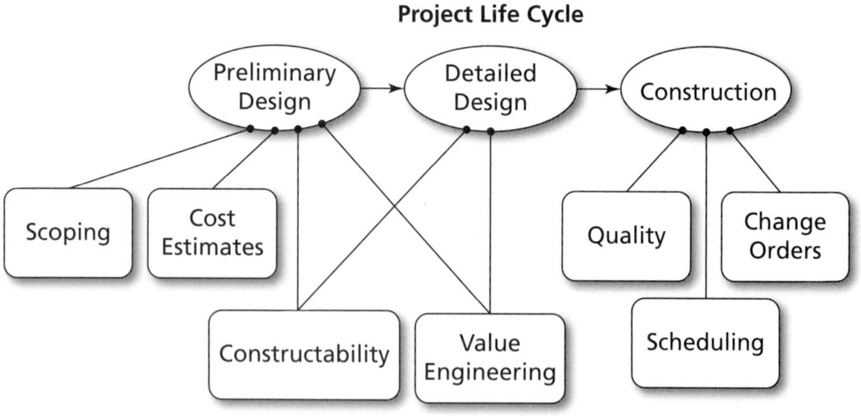

- The CM produces and updates the project schedule as the project proceeds. The CM uses the schedule to coordinate the activities of the various contractors working on the project and sequences the activities of the trade contractors.
- The CM also negotiates change orders with the contractor and coordinates design modifications with the designer.
- Finally, the CM reviews the project during the closeout stage to resolve outstanding issues.

The agency CM provides a broad range of advice to the owner starting from the early stages of a project. These services can help a project run smoothly. Figure 4.3 shows the activities a CM performs throughout the project life cycle to assist the owner. However, as mentioned earlier, an agency CM can make mistakes for which the owner is liable.

Construction Management At-Risk

In CMAR, the CM provides similar pre-design and design services as with agency CM but during construction, the CM becomes the contractor legally responsible for the completion of the project. The CMAR acts as a general contractor during construction and contracts directly with subcontractors to complete the work.

A project can be fast-tracked using this approach. Construction can start before the design is completed. The CMAR and the owner negotiate a **guaranteed maximum price (GMP)** contract on the basis of the partially completed design and the CMAR's estimate of the cost of the unfinished portion of the work. In a GMP contract, the CMAR would be reimbursed for all direct project costs plus a fee up to the negotiated maximum price. If project costs were to exceed this amount, the CM would have to absorb the additional costs. GMP projects often also include incentive clauses that specify that the CMAR will receive additional profit if the project is completed for less than the GMP (Gould, 2005).

Table 4.1 shows the direct costs that typically are reimbursed to the CM during construction. This is important because it defines the allowable project expenses. Table 4.2 shows the reimbursable costs for services that the CMAR performs

Table 4.1 Typical reimbursable direct costs in a CMAR contract

EXPENSE	REIMBURSABLE COSTS
Wages	1. Wages of construction workers directly employed by the CM to perform the construction of the work at the site 2. Actual costs paid or incurred by the CM for taxes, insurance, contributions, assessments, and benefits associated with the construction workers directly employed by the CM and as required by law
Subcontractor costs	Payments made by the CM to trade contractors in accordance with the requirements of the subcontracts
Cost of incorporated materials and equipment	Actual costs, including transportation of materials and equipment incorporated or to be incorporated in the completed construction
Extra materials for wastage and spoilage	Costs of materials in excess of those actually installed but required to provide reasonable allowance for waste and for spoilage
Expendable equipment and hand tools	Actual costs, including transportation, installation, maintenance, dismantling and removal of materials, supplies, temporary facilities, machinery, equipment, and hand tools not customarily owned by the construction workers, which are provided by the CM at the site and fully consumed in the performance of the work
Miscellaneous costs	Laboratory testing, cost of Surveys, utility costs

Table 4.2 CMAR reimbursement for services during construction

EXPENSE	REIMBURSABLE COSTS
Wages	1. Wages or salaries of the CM's supervisory and administrative personnel when stationed at the site 2. Wages or salaries of the CM's supervisory or administrative personnel engaged at factories, workshops (not including CM's principal or branch offices), or on the road expediting the production or transportation of materials or equipment required for the work, but only for that portion of their time required for the work 3. Wages or salaries of the CM's project manager, whether stationed at the site or in the CM's principal office or branch offices, but only for that portion of his or her time required for the work
Materials and equipment/ temporary facilities	Rental charges for temporary facilities, machinery, equipment, and hand tools not customarily owned by the construction workers, which are provided by the CM at the site, whether rented from the CM or others, and costs of transportation, installation, minor repairs and replacements, dismantling, and removal thereof
Miscellaneous	1. That portion of premiums for insurance and bonds directly attributable to the contract 2. Fees and assessments for the building permit and for other permits, licenses, and inspections for which the CM is required by the contract documents to pay

Table 4.3 CMAR costs not reimbursable

EXPENSE	NON-REIMBURSABLE COSTS
Preconstruction services	Any costs incurred by the CM in providing preconstruction services paid under a separate service agreement executed between the CM and the owner
Salaries	Salaries and other compensation of the CM's personnel stationed at the CM's principal office or offices other than the site office, except as specifically provided
Expenses	Expenses of the CM's principal office and offices other than the site office
Overhead	Overhead and general expenses, except as may be expressly included

during construction. This covers the important work of administering, managing, and scheduling the project. The CMAR will have a separate fee-based agreement for preconstruction services. Table 4.3 defines items for which the CMAR will not be reimbursed during the project. CMAR contracts specifically state that the CM will not be reimbursed for any costs exceeding the GMP, nor will the owner be liable for construction errors made by trade contractors hired by the CMAR.

In the context of the basic contract types discussed in Chapter 2, the CMAR, in essence, becomes a contractor with a negotiated contract during the construction phase. It is important to consider the different levels of risk faced by a CM or construction contractor doing negotiated work when compared to those faced by a contractor in competitive bidding. Table 4.2 shows that the CM will be reimbursed for all allowable costs. The CM is not contractually locked into bid prices like a contractor on a competitively bid project. There is less financial risk to the CM or construction contractor performing negotiated work. If the CM negotiates a reasonable GMP and manages the project responsibly, then he or she has a good chance to make a profit.

Case Study: Using CMAR on General Service Administration Projects

The following article describes the use of CMAR in general services administration contracts. Note that the article refers to CMAR as construction manager as constructor (CMc). This article titled "Innovative Federal Agencies Push Home-Grown Methods," was published in *ENR* magazine on March 5, 2007. Tom Nicholson wrote the article.

> It can be said that the bureaucratic halls of federal government agencies traditionally are not places associated with innovative change, especially in today's hyper-paced construction markets where owners in both the public and private sectors must live by the mantra "evolve or die." But the U.S. General Services Administration (GSA) is an exception.
>
> GSA is the procurement arm of the federal government and is responsible for the construction, repair and alteration of all federal buildings. Change at GSA has come in

the form of a gradual break from tradition as project managers increasingly seek new and better ways to deliver fast-tracked, complex construction projects with the speed and efficiency today's markets demand. The traditional project delivery method of design-bid-build, once the sole delivery method used for GSA projects, is beginning to fade as the agency adopts collaborative and integrated delivery.

"In the 1990s, a shift began in how project managers approached project delivery on federal courthouse projects and they began looking at alternative methods," says David Winstead, commissioner of GSA's Pubic Building Service. These days, design-bid-build is obsolete on federal courthouse projects, as project managers exclusively are embracing alternative methods such as construction manager as constructor (CMc) and bridging on high-profile, high-design courthouse projects. While D-B-B is still the most widely used delivery method on GSA's office and border station projects, all of the 31 courthouse projects GSA's Public Building Service currently has in various stages of design and construction are being delivered either by bridging or CMc.

GSA's capital construction program has a total budget of $690 million for new construction, and $866 million for repairs and alteration projects, part of GSA's $10-billion long-range building plan. For this fiscal year, the courthouse budget is $281 million and there is $97 million for site acquisition, design and construction of eight border stations.

Courthouse projects "are becoming increasingly complex and improving the project delivery method to accomplish those projects continues to be a priority," Winstead says. He points to the on-time and on-budget completion last August of the $78-million, 267,000-sq-ft Wayne Lyman Morse U.S. Courthouse in Eugene, Ore., by Kansas City–based J.E. Dunn Construction Co. as an example of GSA's successful use of CMc. "We have had very good results with CMc," he says. "It's becoming a popular delivery method because it brings the builder in early in the project and expedites construction." Winstead also cites the shifting of risk away from the owner through the use of CMc as another reason GSA project managers and executives favor it for courthouse projects.

On the $281-million, 1.5-million sq-ft U.S. Census Bureau headquarters project in Suitland, Md., completed last year, a bridging contract allowed the project team to finish the job about eight months ahead of what would have been possible through D-B-B, says GSA project manager Jag Bhavgava. The project was fast-tracked because of time constraints imposed by the Census Bureau's need to begin preparations for the 2010 census. "It was a short time frame, and the schedule was most critical," Bhavgava says. "The project began in September 2001, and had to be complete by December 2006, and that was a challenge."

The evolution that has taken place in the way the federal government delivers projects has not been a top-to-bottom policy change. Rather, GSA project managers in the field initiated the change. GSA is divided into 11 regions across the country, with project managers and executives in each region responsible for determining the delivery method for each project. The system allows project managers to decide what works best in each region based on a variety of factors such as material costs and labor availability, which often vary by locality, Winstead says.

"Each region has predilection over how projects are delivered," says Bob Fraga, PBS assistant commissioner for capital construction program management. "There are various permutations of CMc being used, sometimes bringing the CM on at the beginning with the architect, and sometimes after design is nearly done."

The CMc method, a home-grown CM at-risk process open to various modifications based on nuances and requirements of each project, was first employed on federal projects in GSA's Southeast region about a decade ago. CMc's Ground Zero was the James H. Quillen U.S. Courthouse in Greenville, Tenn., completed in the mid-1990s.

In the CMc method, the CM is contracted to provide design review, cost estimating, scheduling, establish a guaranteed maximum price and other general construction services. The CMc functions as a member of the GSA project development team and is involved in the planning, design and construction phases of the project. GSA project managers have modified the approach to include "fixed-price-incentive" or "fixed-price-successive-targets" contracts, which establish a "not-to-exceed" price. They place the responsibility and risk on the CMc to deliver the project within the available funds. The contracts also include incentives for the CMc that award a share of any budget savings.

On the Wayne Lyman Morse Courthouse, J.E. Dunn Construction used CMc to finish construction in just over 24 months. Dunn's project executive Ira Gail Wickstrom says the job was originally conceived by GSA as a D-B-B project, but GSA's regional project managers opted to cut time and costs through use of an integrated delivery approach.

"In private work, we have done many CM jobs using a similar approach to CMc called construction manager as general contractor," says Wickstrom. GSA officials expected a 36-month time frame to complete construction, but Wickstrom's experience with CMGC told him it could be done in 24 months. "Had they used [D-B-B], the project would have been a bid buster by $10 million to $12 million," Wickstrom says. "The important part taxpayers should know is that with CMc, they are getting estimates from people who are putting their [butts] on the line." He notes integrated delivery helps the project team shorten schedules, cull unnecessary costs, and navigate mid-stream changes with agility.

"CMc is an evolutionary step up from D-B-B but it is not a silver bullet," Wickstrom says. "Far more important is the attitude the three parties—owner, architect and builder—bring to a project." At the Eugene courthouse, "GSA took ownership. They were more like a private developer that cared about the outcome of the project," he says.

With bridging, a variation of design-build, GSA contracts with a designer to draw up initial designs which "can go as far as 95% into some parts of the project, but only preliminary design on other parts," says Gilbert Delgado, GSA's director of construction excellence and project management. "Then, the CM, who will also be the constructor, takes over completion of the design."

On the Census Bureau job, GSA hired Chicago-based architect Skidmore, Owings and Merrill LLP to do site location, initial design, and the environmental impact statement and assessment. "They did about 25% of the basic concept design," says Bhavgava. "But for the interior design, we wanted to go much further, up to 70%" before hiring a design-build team to finish design and handle construction. GSA awarded Parsippany, N.J.-based Skanska USA Building Inc. the contract to build the complex.

In the bridging method, the handing off of the preliminary design to the design-build team is a critical point in the schedule, a transition stage that can make or break a project's schedule, says Michael Luondi, executive vice president at Skanska USA Building Inc., and the project executive on the Census Bureau project. The handoff was not completely seamless. "The biggest challenge was interpretation of the contract itself," Luondi says. "There were some scuffles. It was a complicated project with a number of components and some parts of the design were far along and others not at all. We probably could have done a better job upfront of making clear what was required, but in the end, it worked out very well."

The Census project reportedly was at the heart of a Skanska reorganization that saw the departure of Michael J. Healy, the chief of the U.S. unit, and the writing off of $47 million in losses. Big hikes in steel prices partly were to blame.

The new approaches to project delivery at GSA have come with some growing pains. The need for contract and project delivery standardization has prompted GSA to develop with the Associated General Contractors a "CMc standard clause," which GSA acting administrator David L. Bibb says will help GSA "package a project a certain way that should fall within a set of expectations regarding the type of information and preparation necessary to prepare a proposal. That's simply more efficient."

GSA also is currently working with the Construction Management Association of America to develop standardized procedures for CMc-delivered projects. "The goal is to develop a national consistency in delivery," Winstead says. He notes GSA project schedules also include peer reviews at the 30%, 60% and 90% stages of completion on every job. GSA has assembled a "select cadre of experienced construction people," who visit jobsites and evaluate progress, Winstead says.

While the evolution in GSA's approach to project delivery is a work-in-progress largely driven by market conditions and the unique requirements of each project, GSA also has initiated in-house programs to help project managers identify and share best practices for project delivery. Project managers from each region meet annually to share and review project delivery experiences. GSA also uses an information-collecting program called the "Project Information Portal" to track GSA's major projects and their respective delivery methods.

Summary

This chapter has described the construction management contract format. Two different methods are commonly used. In agency CM, the construction manager provides advisory services to the owner but does not assume any liability for the conduct of the construction phase. In CMAR, the CM becomes the contractor responsible for the construction of the project and performs the work on a negotiated basis with a GMP. This chapter has provided example contract clauses that illustrate a CMAR's responsibilities during construction. The CM contract format has been found to be useful in shortening project durations through fast-tracking and as a method of providing an unsophisticated owner with construction advice throughout the life cycle of a project.

Key Terms

Agency CM
CM at-risk (CMAR)
Construction manager
Design-Bid-Build
Fast-tracking
Guaranteed maximum price (GMP)

Review Questions

1. Describe how a construction manager is involved during the design of a project.
2. What is the difference between at-risk and agency construction management? Why is CMAR now preferred to agency CM?
3. What is a guaranteed maximum price? Who pays for project costs if it is exceeded?
4. What are the benefits of using the construction manager contract format versus simple competitive bidding?
5. Why is a negotiated contract less risky to a construction manager at-risk than a competitively bid project is to a contractor?
6. What costs can a CMAR be reimbursed for during construction? What costs are typically excluded from reimbursement?
7. Review the case study in this chapter. Why has CMAR become popular with the General Services Administration?

Management Pro

Find a local project using the construction management form of contract. Is it a CMAR or an agency construction management contract? What is the relationship between the owner and CM? How do the CM's activities differ from those of a traditional construction contractor? Is the project fast-tracked? Write a short report describing what you observed about a CM project and how it differs from the traditional method of contracting.

References

Construction Management Association of America. 2002. An owner's guide to construction management. Available from http://cmaanet.org/index.php (accessed May 15, 2007).

Construction Management Association of America. 2007. Choosing the best delivery method for your project. McLean, VA: Construction Management Association of America. Available from http://cmaanet.org/best_delivery_method.php (accessed May 10, 2007).

Gould, Frederick E. 2005. *Managing the Construction Process: Estimating, Scheduling and Project Control.* Upper Saddle River, NJ: Pearson Education.

Halpin, Daniel W. 2006. *Construction Management*, 3rd ed. Hoboken, NJ: John Wiley.

chapter 5

Design-Build: Combining the Designer and the Contractor

Chapter Outline

Introduction
The Design-Bid Process
The Request for Proposals: Defining Project Requirements
Competitive Procurement: One-Step and Two-Step Procurements
Preliminary Designs Included in the RFP
Formation of Teams and Consortiums
Additional Legal Risks for Contractors with Design-Build
Summary
Key Terms
Review Questions
References

Introduction

In a **design-build** contract, a single entity is responsible for designing and building a construction project. This contrasts with the traditional method where there is a separate design contract and a separate construction contract. Design-build projects are constructed by a single company that has both design and construction capabilities or by a consortium consisting of a designer and a contractor. An important aspect of design-build is that the owner has a single party to contact for both design and construction questions, potentially reducing claims and litigation. Figure 5.1 shows the structure of a design-build contract, with the owner contracting with a single entity for both design and construction.

Like the construction management form of contract discussed in Chapter 4, the design-build contract has evolved because of the owner's desire for reduced project durations and knowledge of construction costs beforehand. Although fast-tracking of projects is possible in both design-build and construction management contract types, some owners prefer design-build because it eliminates the need to deal with separate designers and contractors.

Use of the design-build format has been steadily increasing in recent years. In the 1980s, design-build accounted for about 10% of projects in the United States. It is now estimated that over 30% of private construction and 5% of public construction is being built using the design-build contract format. Design-build found its initial application in privately financed commercial buildings. However, there is increasing interest from government agencies in using design-build and in extending its use to heavy construction projects such as highways and mass transportation facilities that government agencies often construct. Many government agencies have been required by law to use only traditional competitive bidding. Many governments are now modifying their laws to allow the use of construction management. Therefore, we should expect to see an increasing percentage of government projects using the design-build contract format.

Design-build is also beginning to be employed for residential construction. Traditionally, architects have designed homes, which were then put out to bid and built by a

Figure 5.1

Design-build contractual relationships

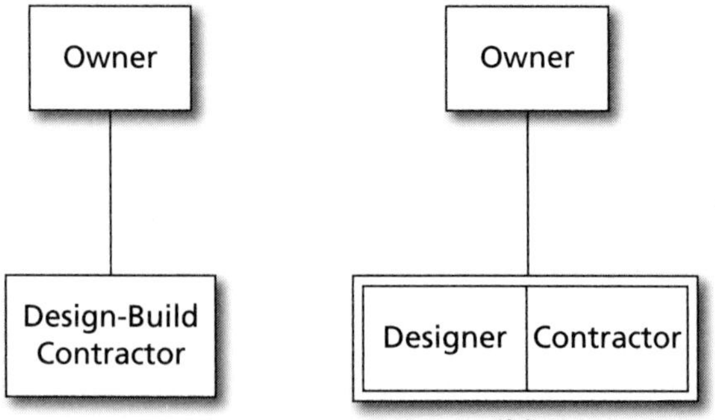

contractor. In residential design-build, a home is built or remodeled and the project is handled by a single company. This method provides a close relationship between an architect and the builders. In the traditional method, an architect may not be consulted on construction issues, and constructors have little input in the design of the home (Barry, 2003).

In other parts of the world also, use of the design-build contract is widespread. For example, 85–95% of all projects in France are conducted using design-build format (Hanscomb, 2004). About 50% of the nonresidential construction projects in the European Community and 70% of the nonresidential construction projects in Japan are constructed using design-build.

Example

The Design-Build Institute of America is an organization that advocates the use of design-build. Its Web site provides information about design-build, industry information, and a database of design-build projects. The Web site can be accessed at http://www.dbia.org/

There are several potential benefits of using the design-build approach (Design-Build Institute of America, 1994):

1. A single entity both designing and constructing a project can fast-track a project more easily and reduce the total project duration.
2. Design-build reduces the adversarial relationships found in traditional design-bid-build construction. The builder and designer are part of the same company and should be working toward a common goal. Questions about the design from the field can be cleared more rapidly.
3. Quality can be improved because the owner documents his or her requirements in terms of the required performance of the facility. The design-builder then warrants to the owner that it will produce a complete design. In traditional design-bid-build construction, the owner provides a complete design to the contractor and then attempts through contract language and inspection to ensure final project quality (Design-Build Institute of America, 1994).
4. Cost savings are possible because design and construction personnel work together to evaluate alternative designs. The design-build methodology allows for extensive value engineering and constructability reviews.
5. The number of change orders due to errors and omissions in the plans is greatly reduced because the design-builder is responsible for producing the design, as opposed to design-bid-build where the owner provides the design to a contractor.

The Design-Bid Process

An important consideration in design-build projects is that the owner can consider qualitative factors such as:

- The design-build team's staff and areas of competence
- The quality of the proposer's work on past projects
- The design-builder's project plan that describes how the project work will be executed
- A preliminary design executed by the design-build proposer if the project is complex

There are several procurement methods that may be used for design-build projects, which are common in the private sector. Using sole source selection, an owner negotiates with a favored set of contractors and then selects the contractor believed to be the most qualified to perform the project. The selection process need not focus on the lowest possible price but can consider qualitative aspects of the competing design-build organizations. In private projects negotiated in this way, guaranteed maximum price (GMP) is often used; however, reimbursement can be by any method including cost-plus a fee and unit price.

Negotiated projects are seen mainly in the private sector. When applying design-build to government-financed projects, federal, state, and local laws typically require some form of competitive procurement. Typically, the competition for government-funded projects is through either a one-step or a two-step procurement process. Both methods allow for consideration of cost and qualitative and technical considerations in selecting the design-builder to construct the project. This is often called **best value procurement** and is based on the idea that the owner may receive the best value by combining consideration of cost and technical factors in selecting among the proposing design-builders.

In competitive design-build procurement for government-financed projects, the proposing design-builders are evaluated on the basis of a weighted average of the cost, qualitative factors such as staffing and experience, and the proposer's technical approach to the project. In two-step design-build, proposers are first screened on the basis of their qualifications, experience, and technical understanding of the project. In the second stage of the process, the short-listed competitors are evaluated on the basis of their detailed plans to execute the project. Both these methods require the owner to objectively score the submitted proposals. Typically, a team of technical experts is assembled to objectively rate the various aspects of the submitted proposal.

The Request for Proposals: Defining Project Requirements

In all methods of design-build procurement, the most important document is the **request for proposals** (RFP). This document is prepared by the owner or consultants to the owner. It is in this document that the owner defines the requirements of the project, and must provide a clear understanding of the scope of work and the required outcomes. For the owner, the development of the RFP is vital to the success of the project. In traditional design-bid-build, the owner provides the contractor with a completed set of plans, and then the contractor must build the project without deviating from the design. In design-build, the owner defines the scope of what is to be constructed in the RFP and leaves the details of the design to the design-builders. This information is essential to the proposing design-builders to develop cost estimates and to insure that the project ultimately meets the owner's requirements.

The other main purpose of the RFP is to define the technical details about the project that the competing design-builders must address in their proposals to show the owner that they understand the nature of the project and to describe how they propose to go about the design and construction. The RFP describes the parts of the design that will require ideas and innovation from the design-build team. For example, on a bridge project, the owner may require the proposers to define their approach to the design of the bridge (perhaps a choice between an arch, cable-stayed, suspension bridge, or some new type of design) with supporting documentation illustrating their design ideas. For a highway project, proposers may be required to include plans for how they will stage construction to minimize disruption to traffic. Development of a good proposal in response to the technical questions posed in an RFP may require considerable work by the proposers. Therefore, owners may often offer stipends or proposal preparation fees to the design-builders to offset some of the proposal preparation costs.

Much like the bid package for design-bid-build construction, the RFP will be used to form the basis of the project contract between the owner and the design-builder. Information included in an RFP typically includes (Design-Build Institute of America, 2007):

1. **General information.** Information about the project and selection procedures
    a. Introduction
    b. RFP schedule
    c. Selection procedure
    d. Selection criteria (and weighting)
    e. Budget (or cost limitations)
    f. Project schedule
2. **Site information.** Qualitative data relating to existing site conditions
    a. Site description
    b. Topographical and boundary survey
    c. Soil investigation data
    d. Utility information
    e. Covenants and restrictions on property

3. **Project requirements.** The project design criteria and description of the expected performance of the facility
   a. Program summary
   b. Functional requirements (goals and objectives)
   c. General physical characteristics/building systems requirements
   d. Performance specifications, including warranties
   e. Codes and standards
   f. Functional relationship diagrams or conceptual building layout
4. **Design-build contract requirements.** May be a summary of contract terms or a copy of the actual contract
   a. Design responsibilities
   b. Construction responsibilities
   c. Responsibilities of the owner
   d. General conditions
   e. Minority participation
5. **Requirements for proposal.** A description of the information that proposers must include in their proposal
   a. Drawings
   b. Specifications
   c. Design-build organization
   d. Project personnel
   e. Quality control program
   f. Schedule
   g. Price proposal
   h. Proposals for solving design and construction issues

Figure 5.2 shows portions of a standard RFP used by the Naval Facilities Engineering Command (1996). This part of the RFP describes a two-step evaluation procedure and provides details to the proposers about the evaluation criteria. Note that there is a space for the Navy personnel to fill in a description of the scope of work to be performed and to provide a preliminary estimate of the project cost. Examination of the submission requirements indicates that the proposer's technical solutions to the design and construction of the project become an important factor in the second stage of the evaluation process. This material from the standard Naval Facilities Engineering Command (NAVFAC) RFP also clearly states that best value procurement will be used, and that neither the low bid nor the best technical proposal will necessarily be selected. When preparing the RFP, Navy personnel would add specific design issues that must be addressed by the proposers. The RFP also provides a description of how NAVFAC will evaluate and rate the submitted proposals. This is illustrated in Figure 5.3 (Naval Facilities Engineering Command, 1996).

1.1 GENERAL CONTRACT DESCRIPTION

[Insert brief contract description including scope and estimated cost range]

1.2 GENERAL OVERVIEW OF THE PROCUREMENT PROCESS

[Insert brief description of the two-Phase design-build source selection procedures for this Request for Proposal (RFP)]

The Government reserves the right to reject any or all proposals at any time prior to selection in Phase I. Offerors are advised that selection in Phase I may be made without discussion or any contact concerning the proposals received in Phase I. Therefore, Offerors shall not assume they will be contacted or afforded an opportunity to qualify, discuss, or revise their Phase I technical proposals. However, the Government reserves the right to clarify certain aspects of proposals or conduct discussions providing an opportunity for the Offeror to revise its proposal. In Phase II, the Government also reserves the right to reject any or all proposals at any time prior to award; to negotiate with any or all Offerors; to use a tradeoff process when it may be in the best interest of the Government to consider award to other than the lowest priced Offeror or other than the highest technically rated Offeror; and award to the Offeror submitting the proposal determined by the Government to be the most advantageous (Best Value) to the Government. Offerors are advised that an award may be made in Phase II without discussion or any contact concerning the proposals received. Offerors shall not assume that they will be contacted or afforded an opportunity to qualify, discuss or revise their proposals. However, the Government reserves the right to clarify certain aspects of proposals or conduct discussions providing an opportunity for the Offeror to revise its proposal.

PHASE I EVALUATION FACTORS: Offerors will be evaluated on the following Factors that are of equal significance: Subfactors under each Factor are of equal significance.

Factor A—Past Performance

1. Design Team
2. Construction Team

Factor B—Small Business Subcontracting Effort

Factor C—Technical Qualifications

1. Design Team
2. Construction Team

Factor D—Management Approach

[Mandatory Subfactor 1—Safety]

PHASE II EVALUATION FACTORS: A maximum as many as five (5) Offerors will advance to Phase II. Phase II will be evaluated on the Offeror's technical proposal and price proposal that are approximately equal in significance. Offerors will be required to submit separate technical and price proposals.

(Provided for information only)—Offerors are not required to submit this information until Phase II. Specific Phase II submittal criteria will be issued as an amendment to this solicitation to those Offerors selected in Phase I.

Award will be made to the Offeror whose technical proposal and total evaluated price provide the best value to the Government using a tradeoff process, price and other factors considered. All evaluation factors other than price, when combined, are approximately equal in significance to price.

It is anticipated that the following equally significant technical evaluation factors will be included in Phase II:

Factor A—Past Performance [same as Phase I unless conditions change]

Factor B—Small Business Subcontracting effort

Factor C—Technical Qualifications [same as Phase I unless conditions change]

Factor D—Technical Solution

Figure 5.2 Portion of NAVFAC RFP describing proposal evaluation method

EXCEPTIONAL (E): The proposal demonstrates a thorough and detailed understanding of the requirements and needs the stated requirements of the RFP. Technical approach and capabilities significantly exceed performance and capability standards. Proposal/factor offers one or more strengths. Strengths significantly outweigh weaknesses, if any. The proposal/factor represents a high probability of success with no apparent risk in meeting the Government's requirements.

GOOD (G): Proposal/factor demonstrates a good understanding of the requirements. Technical approach and capabilities exceed performance and capability standards. Proposal/factor offers one or more strengths. Strengths outweigh any weaknesses. The proposal/factor represents a strong probability of success with overall low degree of risk in meeting the Government's requirements.

ACCEPTABLE (A): Proposal/factor demonstrates an acceptable approach that meets the stated RFP requirements. There is little, or a manageable amount of risk of failure to meet the RFP requirements. There is a general understanding of the requirements. Proposal/factor offers no strengths, or, if there are any strengths are offset by weaknesses.

MARGINAL (M): Proposal/factor demonstrates a limited understanding of the RFP requirements. Technical approach and capabilities are questionable as to whether or not they meet performance and capability standards necessary for acceptable contract performance. Proposal/factor contains weaknesses and offers no strengths, or if there is any strength, these strengths are outweighed by the weaknesses. The proposal/factor represents a low probability of success with overall high degree of risk in meeting the Government's requirements. Proposal/factor might be made satisfactory with considerable additional information and without a major revision of the proposal.

POOR (P): Proposal/factor demonstrates a lack of understanding of the RFP requirements. Technical approach and capabilities do not meet performance and capability standards necessary for acceptable contract performance. Proposal/factor contains major errors, omissions, significant weaknesses and/or deficiencies. The proposal/factor represents a very low probability of success with an extremely high degree of risk in meeting the Government's requirements. Proposal/factor could only be made satisfactory with a major revision of the proposal.

[The rating of plus (+) or minus (–) will be used in order to differentiate the strengths and weaknesses of an Offeror. Note that the narrative can be used to differentiate between high and low ends of a rating.]

Figure 5.3 Portion of NAVFAC RFP describing proposal rating system

The complete text of the Naval Facilities Command RFP for design-build projects can be found at the NAVFAC design-build request for proposal Web site at http://www.wbdg.org/ndbm/

Competitive Procurement: One-Step and Two-Step Procurements

In this section, more details about one- and two-step procurement processes are presented. Figures 5.1 and 5.2 show that NAVFAC uses a two-step procurement process and that these procedures are carefully spelled out to proposers in the RFP. Figure 5.4 shows the one-step procurement process. In the one-step process, an RFP is advertised by the owner. This RFP will request that the design-builder's proposal include a statement of qualifications (SOQ), cost estimate, and technical description of the approach to the project. The RFP is examined by the competing design-builders who submit proposals that are evaluated by technical experts employed by the owner. The highest scoring proposal is selected to perform the project.

In the two-stage method, proposers are initially screened on their qualifications to conduct the project. A request for qualifications (RFQ) and a preliminary RFP are

Figure 5.4

One-step design-build procurement

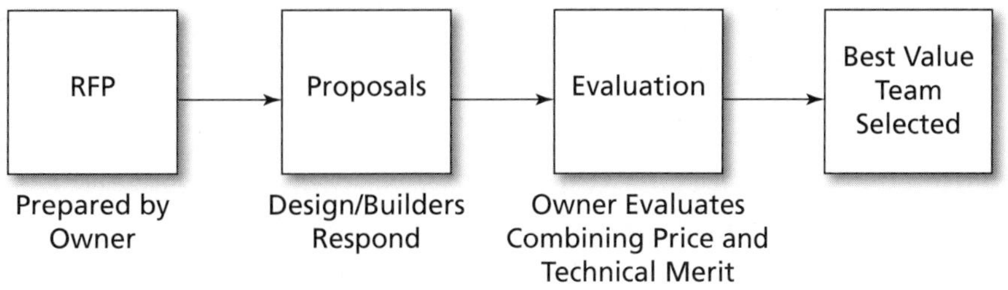

advertised by the owner. The RFQ typically requests detailed information about a contractor's experience and performance on similar projects. Figure 5.5 shows the flow of paperwork for a two-stage evaluation. Figure 5.2 has already shown the information that NAVFAC requires design-builders to submit with their proposal. However, NAVFAC performs many projects that involve the building or retrofitting of buildings. Many state highway agencies are starting to use design-build for complex highway infrastructure projects. Their requirements for information about the design-build proposers in the first step can be extensive. An example of some of the information that the Washington State Department of Transportation requests is (Washington DOT manual; see Washington State Department of Transportation, 2004):

- Individual experience of team members with design-build contracting
- Corporate experience with design-build contracting
- Specialized design capabilities that relate to the project
- Experience with complex construction staging
- Safety record
- Quality assurance and quality control procedures
- Understanding of local environment and sensitive community issues

Figure 5.5

Two-step design-build procurement

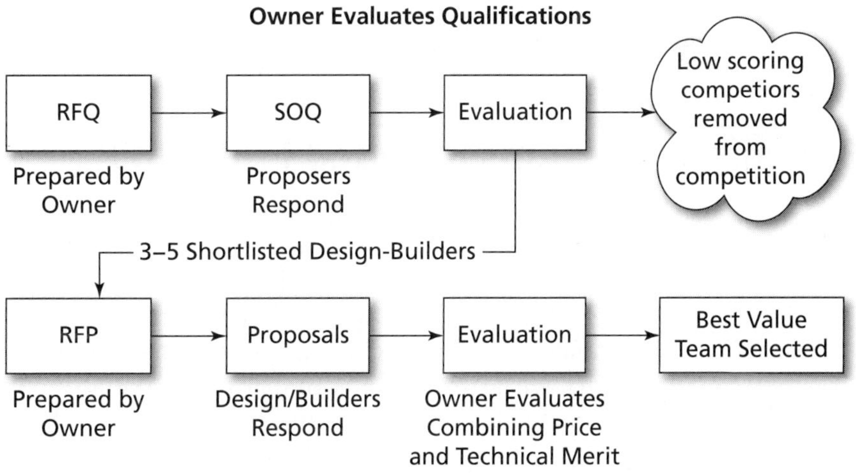

- Scheduling and project control systems used by the proposer
- Specialized expertise the proposer may have that applies to the project

On the basis of the information requested by the owner, the design-builders develop their SOQ. The owner must evaluate all the submitted SOQ to determine a "shortlist" of what will be provided with a final version of the RFP and asked to submit a proposal. The SOQ is evaluated by a team of technical experts in the owner's organization. After selecting the short-listed firms, this same team will score the submitted proposals. The submitted proposal will contain the proposers' detailed discussions of their approach to the design and construction issues of the project. The evaluating team will consist of experts from within the owner's organization with expertise in the various design and construction types included in the project. For example, if a project includes a great deal of pile-driving work, then geotechnical experts will be part of the evaluation team so they can analyze the contractors proposed pile-driving schemes. In the second step of the process, the short-listed firms or consortiums are evaluated on the basis of the cost and technical merit of their proposals (Figure 5.2 provides the NAVFACS evaluation criteria). The firm with the highest number of points wins the competition. The winning competitor may not be the competitor offering the lowest price. Depending on the owner, the weighting of price and technical merit can vary widely. For highway projects in Great Britain, a weighting of 60% quality and 40% cost is often used, whereas in Sweden, 70% cost and 30% technical factors is widely used (Cox et al., 2002). The ultimate goal is to produce a project that provides the best overall value to the owner.

One of the potential disadvantages of the using best value procurement is that the owner's staff must assign points to various aspects of a proposal. There may be no completely objective method to award points to various elements of the proposal. For example, there may be no objective way to state that one firm's construction staging plan merits 40 points while another firm's plan merits 43. Comparison of the value added by alternative proposals may also be difficult. If one competitor includes higher quality piping in his or her proposal, it may be difficult to compare the added benefit with another proposer who has included an advanced electrical system in his or her design (Hill, 2005).

Preliminary Designs Included in the RFP

The percentage of the design that must be prepared before a design-build RFP can be issued varies according to the project type. On some projects such as a building, no design may be necessary and only some preliminary drawings are included in the RFP as an aid in defining the scope and requirements. However, a transportation project such as a new light rail line may have additional complications. The primary difficulty with a rail line is that it may take a wide variety of routes between two points in a city. The owner may require the **right of way** to pass through certain neighborhoods or may even wish to specify the route exactly. In either event, the owner may include a preliminary design in the RFP that must be used as a starting point by the design-build firm. Typically, the owner will complete this preliminary design or hire a separate design consultant to prepare the preliminary design. If 10–20% of the design is complete, it can be considered a design criteria design-build project. If more than 20% of the design is complete, then a project can be considered a preliminary design design-build project.

Formation of Teams and Consortiums

In some types of construction, such as large petrochemical plants, large firms exist that have in-house design and construction capabilities. However, for many types of construction, a designer and a builder will need to team together for a project. Other specialty firms may also be needed on a project team. Legal teaming agreements must be signed. The teaming agreement can take several forms, with one party being the lead contractor with the other team members having a subcontractor relationship with the prime, or a joint venture can be formed among the team members. A limited liability corporation is often used to form joint ventures (The Associated General Contractors of America, 1999).

A successful design-build project must team companies that have similar cultures. Clear communication and problem-solving techniques should be set up. Formal partnering (see Chapter 3) is often useful to promote cooperation between designers and builders. Tremendous benefits can accrue from the design-build process by integrating the knowledge of the construction company with that of the designers. There are excellent opportunities to employ value engineering to the project design by using the contractor's personnel to provide input to the design.

Additional Legal Risks for Contractors with Design-Build

In traditional design-bid-build contracting, the owner provides the contractor with a completed set of plans. The owner warrants that the provided plans are accurate and suitable for the project. Courts have ruled that if the contractor performs the work in compliance with the owner-specified design, the contractor cannot be held responsible if the design is defective (The Associated General Contractors of America, 1999). However, using a design-build contract format, the design-builder is responsible for both design and construction. The design-builder assumes responsibility for the plans and warrants that the plans and specifications are accurate and suitable for the project's intended use. The design-builder becomes responsible for insuring that the design conforms to all applicable regulations. This additional liability requires that construction contractors involved in design-build projects must carry special coverage for design defects errors and omissions.

Case Study: A Successful Design-Build Project

The following article excerpt describes the construction of a baseball stadium in Washington, DC. It shows how fast-tracking and design-build can reduce project duration. This article titled "Baseball Park in Nation's Capital is on its Way to Break the Speed Record," was published in ENR Magazine on December 5, 2007. Nadine M. Post wrote the article.

> With only 23 months to complete all the bases, the team building the 85%-complete D.C. Major League baseball park is getting very close to hitting construction's equivalent of a grand slam off a 100-mph pitch. If it opens April 1 as planned, the $611-million home for the Washington Nationals will break the speed record for major-league ballpark construction.
>
> "We're probably pushing the limits of fast track," says Rick "Buck" Buckovich, senior project manager for structural steel and precast for Clark/Hunt/Smoot (CHS). The Bethesda, Md.-based design-build joint venture is led by Clark Construction Group and includes Hunt Construction Group and Smoot Construction.
>
> Despite fast-track's drawbacks, which include redoing some work, "we felt we could do [this job] from the day we "signed the contract," says Ron Strompf, CHS's senior project superintendent.
>
> The job is not only breaking records, it is making them. The 41,000-seat ballpark is likely to garner the distinction of being the first major league sports facility to achieve certification from the U.S. Green Building Council's Leadership in Energy and Environmental Design (LEED) rating system.
>
> Design-build, with preconstruction services and design-assist from major subcontractors, allowed CHS to break the ballpark into nine discrete, sequenced segments for purposes of both design and construction. The strategy saved six months in construction, say CHS executives. "It's not unique to break a building into segments but it is unusual to design those segments in the construction sequence and manage to start construction earlier," says Alan Petrasek, a CHS project executive.
>
> That put stress on the architect. "We were in schematics and they were building it, which is not an optimum arrangement," says Susan Klumpp, project manager for the local joint venture architect HOK/Devrouax & Purnell Architects (HOK/DP). "It was really a hold-onto-your-hat pace," she says. Even in the interior fit-out stage, "it's still tough," she adds.
>
> But because of CHS's strategy, steel fabrication and erection took a year less than is customary, says Don Banker, president and CEO of Banker Steel Co. LLC, Lynchburg, Va. "Clark did a great job 'policing' the collaborative process," says Banker, who explains he has seen the strategy fail with other design-build contractors because of weak leadership. "Thanks to Buck, everyone had their marching orders when they left the meetings," says Banker.
>
> The design-assist process "brought everybody's issues to the table in real time," adds Mark J. Tamaro, project manager for the local structural engineer, Restl/Thornton Tomasetti Joint Venture. "We could use the contractors' skills to optimize system selection based on schedule instead of speculating or making assumptions about construction."

Key to the process were weekly structural design meetings, which started in the preconstruction phase. Everyone from the steel detailer to structural precaster R.W. Sidley, Thompson, Ohio, put their heads together at the meetings to determine the best way to expedite the schedule while keeping within the budget. "It was great," says Versie Stephenson, structural coordinator for HOK/PD. "Usually, we get through CDs before the subs are on board," he says.

A 3D digital model for the structural steel helped the team, cutting two months off the project's timeline and decrease structural requests for information to about 10% of what is usual, according to team leaders. "The bottom line is that the job could not have been completed for the 2008 season using traditional methods," says Matthew T. Haas, a CHS project executive.

Thornton Tomasetti pioneered the use of a digital model on Chicago's Soldier Field football stadium, which opened in 2003. The ballpark job, TT's second building information model (BIM) for a sports job, was a much better experience because all team members supported its use, says the engineer.

Using the model at the design-assist meetings, the team worked out complex geometries of exposed connections to satisfy the architect's aesthetic requirements and the engineer's safety requirements. The steel detailer's favorite part of the work was talking directly to the architect, something that rarely happens. "We had long meetings about [variations in] slab thickness" and its impact on the position of the steel supports, says steel detailer John Shaw, president of Sharpsville, Md.- based Mountain Enterprises.

Beginning

Ballpark design started in May 2005 for owner D.C. Sports & Entertainment Commission. CHS, which was selected in August 2005 for preconstruction work, signed its contract with DCSEC in March 2006 after submitting a $344-million guaranteed maximum price in late January 2006. That was at the end of schematic design and nine months earlier than anticipated, says CHS.

Based on CHS's guaranteed maximum price, the District of Columbia funded the project in February 2006. The ballpark's revenue bonds, secured by sales tax from the ballpark, other tax revenue and the team's annual rent, were sold in May 2006. Altogether, the district borrowed $534.8 million.

DCSEC took control of the site in March 2006. That also marked the beginning of demolition work on existing buildings and remediation of the brownfield site, near the Anacostia River. Some 337,000 tons of contaminated soil have been removed.

Workers began excavation on May 4, 2006, just a couple of days after the Lerner Group bought the team from the owners of major league baseball. Steel erection started on Oct. 5, 2006—one day early. Steel was topped out in July. The guaranteed maximum price has now grown to $388 million because of changes made by the team but the job is on budget and on schedule for its completion in April, say CHS executives.

The joint design decisions resulted in a hybrid frame—structural concrete on the lower third because there was no lead time and concrete and other materials were readily available, and structural steel above because of the longer lead time and speed of erection. The main concourse level is the transition floor, with steel columns encased in concrete.

To keep to the schedule, it was critical to claim a place in line for mill rollings eight months in advance of the start of steel erection that was set for Oct. 6, 2006. To expedite construction, DCSEC authorized and paid for the steel mill reservation in January 2006. Also, CHS started preconstruction meetings with the design team and major subcontractors in late 2005.

Creating a digital model for the structural steel made the mill reservation tonnage less of a rough estimate. "We possibly could have done the estimate without the model but it would have been highly inaccurate," says Tamaro.

For the reservation, the engineer designed the first of seven major sectors of steel. It then extrapolated the total tonnage. That was risky, considering ballparks are asymmetrical. "We went back and forth" with Mountain Enterprises during the process, says Tamaro.

The pressure to reserve the mill forced the architect to lock the geometry of the ballpark six months earlier than usual. The engineer needed the information to create the model, says HOK/DP's Stephenson.

On Feb. 2, 2006, Banker received a fax from the structural engineer with the total tonnage of 8,000 tons. Banker made the mill reservation soon thereafter.

Because design was not complete, Banker's price was based on unit prices and quantities rather than a hard bid. As the design progressed, the numbers were converted to a contract value.

"We were partially at risk and [that's why we] wanted a seat at the table, to make sure the design matched up with the unit price values," says Banker.

"Banker and Clark stuck their necks out," adds Tamaro. "That was pretty gutsy. It was a tough time."

It turns out the estimate was pretty good. At the end of the project there is no leftover material, says Banker. "During the course of the project there were materials that were extra due to changing conditions, but it was all consumed by job completion," he says.

The next pressure after the mill reservation was the five mill orders. The engineer designed each sector, following the construction sequence from the right field side and moving clockwise.

At a certain stage of detail to meet the mill order schedule, each sector model would go to the fabricator for the release of the mill order. Then, the model went to the steel detailer, who checked all the positions of the steel and detailed the connections, finishing the model. Finally, the model moved to the fabricator for production.

"Typically, we get all the drawings at once and [still] create similar sectors," says Banker. "The problem with that is we've waited six months to a year to get the whole design, plus there is no opportunity during design for input from the fabricator and erector."

Paper shop drawings were produced from the model. Many of them were reviewed by the engineer before it issued final paper design documents. Almost like as-built drawings, the paper documents were intended to match the model, not the other way around. The unit-price arrangement reduced the reliance on paper documents, says Tamaro. "The final model allowed everyone to know exactly what the steel quantities were," he says.

During this time, construction was proceeding in a "wedding cake" sequence. Foundations and concrete framing progressed around the bowl segment by segment, followed by a level of steel framing that was followed by precast seating. The pattern continued until the building topped out. The trades led each other around the bowl by as much as a dozen bays or bents to as few as two.

To maintain schedule, steel erection had to be completed by July. The goal was to get all elements, including precast seating risers, installed so the field area could be cleared of cranes in time to excavate for drainage systems. Sod installation had to be completed by November.

The baseball team made changes last March, including calling for a more elaborate scoreboard. The change almost threw a wrench in the works, says CHS. To expedite the process, the structural engineer worked with the fabricator to redesign the scoreboard using steel sections and shapes already owned by the project. Scoreboard erection had to be resequenced by Banker, CHS and Bosworth Steel Erectors Inc., Dallas, "to make it the last pick on the job," says Tamaro.

The sod now is in but the schedule remains. Electricians are running cable, plumbers are hooking up concessions, washrooms and more, and workers are installing about 1,800 seats each week. In addition, perimeter sitework is ongoing, as well as glazing and interior finish work.

There are currently 1,185 requests for information on the project. CHS officials say 10,000 requests would be more typical for a project of this magnitude. In addition, there are no major outstanding change orders or claims, CHS reports.

"I'm nervous but I know we are going to make it," says CHS's Strompf. "We've developed a head of steam and nothing can stop us."

Summary

The design-build project format is gaining rapidly in popularity in the United States and the rest of the world, primarily because of owners' dissatisfaction with the traditional design-bid-build process. A single company or team of companies is responsible for both design and construction of the project. Design-build was initially used for privately financed commercial buildings, and the design-build team was selected through negotiation. There has been increasing use of competitive methods to select design-builders. The one- and two-step methods allow for consideration of factors other than the lowest price in the selection process. These competitive procurement methods have allowed government agencies to increasingly employ design-build.

In this and the preceding chapters, we have seen different methods of construction procurement. Although design-build has been increasing in popularity, it may not be appropriate for all projects. Owners must carefully select the type of construction contract to use on the basis of the project's characteristics. In Chapter 6, new and evolving methods of contract procurement will be discussed that extend the original design-build concept by involving the design-builder in more areas of the project life cycle.

Key Terms

Best value procurement
Design-build
Request for proposals (RFP)
Right of Way

Review Questions

1. The use of design-build projects has been decreasing. T or F.
2. Design-build eliminates the need for an owner to deal with a separate designer and contractor. T or F.
3. Discuss the potential benefits of using design-build on a project.
4. What is the request for proposal? Why is it important?
5. What is the difference between one-step and two-step evaluation of design-build projects?
6. Figure 5.2 shows the requirements for a two-step evaluation process. How is the selection of the design-builder described in this figure different from the selection of a contractor using traditional low bidding?

Management Pro

Make a table of the advantages and disadvantages of traditional contracts, agency CM, CMAR, and design-build. What type of contract would you prefer to use if you were a contractor? Why?

References

Barry, Tom. 2003. Design-build construction enters residential market. *Atlanta Business Journal* November 11. Available from http://atlanta.bizjournals.com/atlanta/stories/2003/11/03/focus3.html (accessed July 22, 2007).

Cox, David O., Keith R. Molenaar, James J. Ernzen, Gregory Henk, Tanya C. Matthews, Nancy Smith, Ronald C. Williams, Frank Gee, Jeffrey Kolb, Len Sanderson, Gary C. Whited, John W. Wight, and Gerald Yakowenko. 2002. Contract administration: Technology and practice in Europe, report no. FHWA-PL-02-0xx. Washington, DC: Federal Highway Administration.

Design-Build Institute of America. 1994. An introduction to design-build. Available from http://www.dbia.org (accessed June 1, 2007).

Design-Build Institute of America. 2007. The design/build process—Utilizing competitive selection. Available from http://www.dbia.org (accessed July 30, 2007).

Hanscomb. 2004. Design-build becoming a revolution. Hanscomb means report 16, no. 4. Available from http://www.icoste.org/roundup1204/RoundupOct04ChairmansMessageNews.html#hanscomb (accessed July 15, 2007).

Hill, Elizabeth G. 2005. Design-build: An alternative construction system. Sacramento, CA: Legislative Analyst's Office. Available from http://www.lao.ca.gov/2005/design_build/design_build_020305.pdf (accessed July 15, 2007).

Naval Facilities Engineering Command. 1996. Evaluation factors for award, NAVFAC design-build request for proposals, document 00210. Available from http://www.wbdg.org/ndbm/Division00/Div00_00210_FINAL.html?Section=00210N (accessed July 25, 2007).

The Associated General Contractors of America. 1999. Construction contractor's guide to the design-build process, report 2910. Arlington, VA: The Associated General Contractors of America.

Washington State Department of Transportation. 2004. Guidebook for design-build highway project development. Available from http://www.wsdot.wa.gov/Projects/delivery/designbuild/default.htm (accessed July 10, 2007).

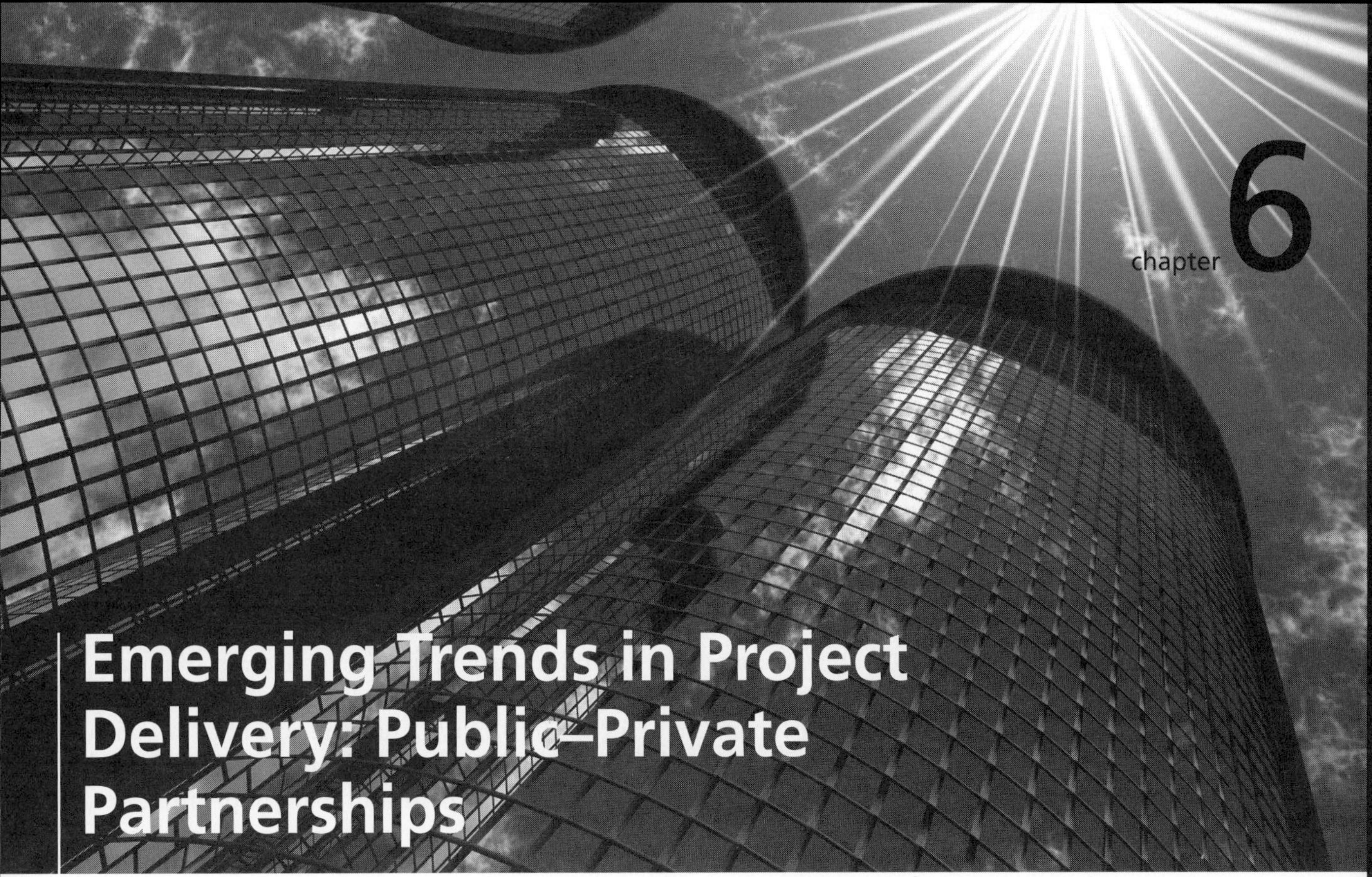

chapter 6

Emerging Trends in Project Delivery: Public–Private Partnerships

Chapter Outline

Introduction

In Chapter 5, on design-build procurement, we discussed the emergence of a contract format that promotes cooperation and reduces friction between the owner and the private parties that design and build projects. **Public–private partnerships (PPPs)** are emerging as an extension of the design-build concept to include the private sector in the whole life cycle of public works projects, including the financing of the project. Public–private partnerships are defined by the National Highway Institute as "an arrangement of roles and relationships in which two or more public and private entities coordinate/combine complementary resources to achieve their separate objectives through joint pursuit of one or more common objectives" (Lawther, 2002). PPPs are a way for government agencies lacking the financial resources to satisfy the demand for new facilities in traditional ways that can rehabilitate and expand existing infrastructure.

PPPs typically involve the use of private capital to design, finance, construct, maintain, and operate a project for public use for a specific time period during which a private consortium collects revenues from the users of the facility. When the consortium's term expires, title to the project reverts to the government. By then, the consortium should have collected enough revenue to recapture its investment and make a profit (Levy, 1996).

Reasons for the Emergence of Public–Private Partnerships

There are several reasons for the current interest in PPPs. One of them is greater efficiency in the use of public resources. Experience in the use of PPP in Europe has shown that the private sector can be more efficient than government agencies in delivering infrastructure projects. It has been estimated that state and local governments experience cost savings of 10–40% through the use of PPP privatization schemes (National Council for Public–Private Partnerships, 2002). Additionally, PPPs are a means of increasing investment in infrastructure. We discussed in Chapter 1 how infrastructure in the United States has deteriorated and is given a low score in the ASCE's annual report card including bridges, transportation facilities, water supply systems, and wastewater treatment. Economic growth is highly dependent on the enhancement and development of a nation's infrastructure. There is also an urgent need for new social infrastructure such as hospitals, prisons, educational facilities, and housing. As populations grow, there is increasing demand for all types of infrastructure, and government agencies may not have the financial resources to cope with the demand alone (Middleton, 2001).

Partnerships between government and the private sector address government needs in several ways (Savas, 2000):

- Private companies help governments to identify new user-financed, profit-making facilities or existing facilities in need of renovation or expansion. Private, profit-oriented businesses have a financial incentive to seek out new projects that would otherwise go unbuilt until government funds became available.

- Involvement of private sponsors and experienced commercial lenders assures in-depth review of the technical and financial feasibility of a project because of the need to assure profitability.
- The private sector can access private capital markets to supplement or substitute for scarce government funds.
- Usually, the private sector builds more quickly and more cost effectively than a government can. Construction is generally more rapid because private developers are more flexible and do not have to observe government procurement rules and there is less bureaucracy to delay project schedules.

Potentially, the use of PPPs will allow government agencies to build more projects more efficiently. A PPP can be structured in various ways. In the following section, we will discuss the different types of PPPs.

The Structure of Public–Private Partnerships

Various types of PPPs are in use around the world. Different variations in methods are possible depending on the situation and the level of privatization that is required. Several of the primary methods including build-operate-transfer and design-build-finance-operate will be discussed in this section.

Build-Operate-Transfer

Build-operate-transfer (BOT) is a PPP method that is often used. Under this type of contract, a concession is granted to a consortium to design, finance, operate, and maintain a facility for a period, usually between 10 and 30 years. This is usually applied to large infrastructure projects such as highways, tunnels, and bridges. The contractor recovers the cost of the project plus profit by collecting tolls during the life of the concession period. Typically, at the end of the operating period, all operating rights and maintenance responsibilities revert to the government.

BOT projects are frequently used in developing countries as a means to obtain funds for the much needed infrastructure projects. The types of projects funded are diverse. For example, the Philippines has undertaken BOT projects for shipping terminals, telecommunications, power generation, and industrial parks. With increased urbanization, developing areas require significant infrastructure investments. It is anticipated that much of this investment can be in the form of PPPs using a BOT form of contract.

PPPs on large projects require consortiums of designers, construction contractors, financiers, and other disciplines to form a concession company. The arrangements can be complex, and there is no fixed structure for concession companies or the form of contractual obligations between parties. Figure 6.1 shows the structure of a BOT and illustrates the different parties that are contractually related. A bidding consortium of companies owns the concession company. The concession company has a concession agreement with the host government agency that allows the contractor to take control of the facility for a given period of time. Loan agreements are obtained from

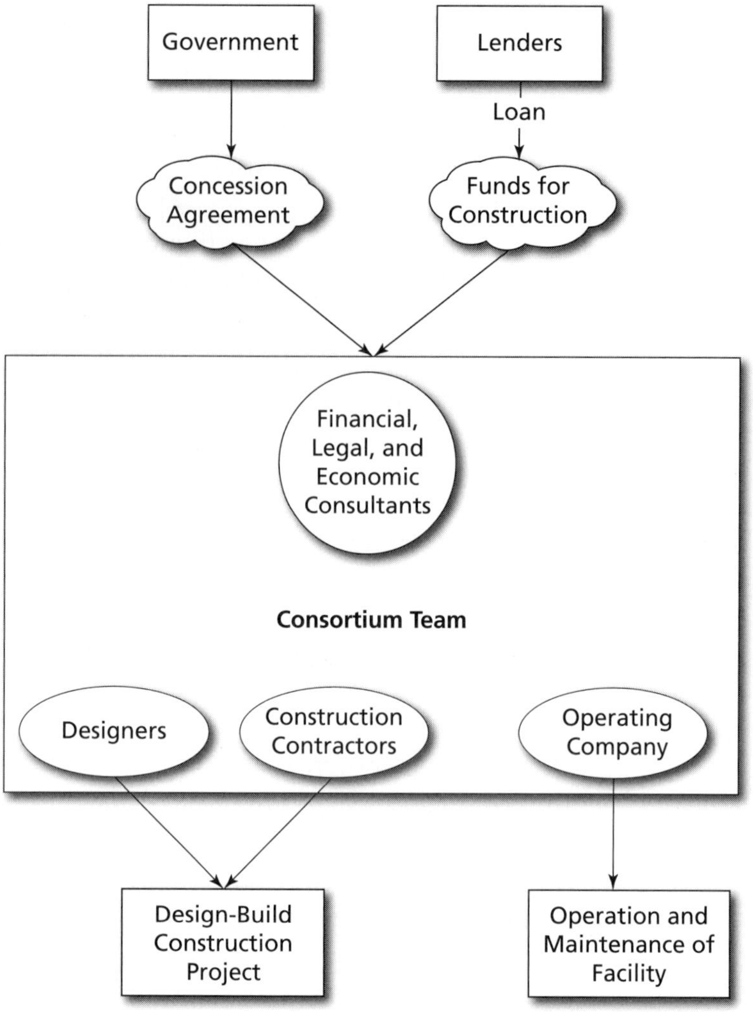

Figure 6.1

Structure of a public–private partnership consortium

various debt providers to finance the project. In the construction phase of the project, the concession company has a contract with joint venture construction companies to construct the project. Finally, the concession company has a contract with an operating company (often the same construction company that built the project) to operate and manage the facility during the concession period.

In BOT projects, the document that defines the relationship between the government agency and the consortium building the project is the concession agreement. It establishes the concession rules and the contractual rights of the main parties. The principal issues that are dealt with in a concession agreement include:

- The nature and length of the concession, scope of the work, and operation of the completed facility
- A specification of what is to be provided

- The extent of permitted variations to the specification
- The performance standards to be achieved
- The tolls, prices, or payments to be charged, together with any arrangements for adjustments
- Provisions to ensure the concessionaire's rights in the event of changes to any enabling legislation and any payments that might accrue therefrom
- Provisions for the termination of the contract
- The circumstances in which the grantor of the concession will be permitted to take over the concession, and the rights of the parties should this occur before the end of the concession period (Smith, 1999)

A successful concession requires a feeling of partnership between the government and the consortium partners. Five factors have been identified that appear to be necessary for each major participant in a BOT project to have the maximum chance of achieving their goals (Smith, 1999). First, there must be a genuine desire for a win-win solution with common agreement among the parties as to their mutual and individual objectives. A BOT approach requires more teamwork than conventional contract types. Second, a complex BOT requires a strong, persistent, and persuasive project leader to fight for the project. Third, there should be adequate and accurate data to conduct a financial risk assessment of both the procurement and operational phases, with responsibility for managing the risks placed with the party best able to control them. Fourth, an accurate calculation of the project's economics is necessary, including length of concession, and assessments of the influence on income and expenditure of project risks. Finally, choice of the correct procurement methodology is important for the construction phase. Consortium contractors often perform the construction in a design-build format.

BOT projects have several advantages for the host government agency and its citizens (Levy, 1996):

- Infrastructure projects can be built at little cost to the taxpayer.
- The private sector can usually move preconstruction and construction along more rapidly than government agencies, allowing projects to be completed more quickly.
- The sponsors must operate and maintain the facility for a period of time exceeding 20 years. Therefore, the initial construction quality of the facility will be high to reduce the BOT consortiums long-term maintenance costs.
- Taxes will not have to be increased, nor will revenue bonds need to be sold to finance the construction of the project.

Clearly, the BOT method has many advantages for a government owner. However, construction of a BOT project requires the formation of a team of organizations that is much more sophisticated than traditional construction arrangements. A construction contractor would need to work not only with a designer but also with bankers, investment firms, and economic consultants to build a BOT project.

Members of a BOT consortium can make substantial profits from their participation in an infrastructure project. However, there are several risks that can occur, particularly for international projects in developing countries. Political instability in the host country is a concern at all stages of a BOT project. Because most concessions are from 20 to 40 years, long-term political stability is important. There is also the risk of significant cost overruns on a BOT project that may change a project's economic viability. If additional financing is not available, the project can come to a halt or end in default. There is also the risk of unfavorable currency devaluations that can cause a BOT consortium to pay back loans with devalued revenue. Another risk is the level of the usage fee set for a facility. Toll rates for concession-type highways built in Mexico had tolls about eight times higher than comparable tolls in the United States. This resulted in increased toll jumping and reduced revenue for the consortium that built the project. Finally, erroneous projections of facility use over the concessionary period may substantially affect revenue. A BOT consortium's source of revenue is based on projections of the number of consumers who will use a facility. If the number of consumers is less than projected, it can have disastrous effects on the profitability of the BOT venture (Levy, 1996).

Design-Build-Finance-Operate

The **design-build-finance-operate (DBFO)** type of project is similar to the BOT format. However, it is different in one important aspect. A DBFO consortium is responsible for the design, construction, maintenance, and operation of a facility. The DBFO consortium also finances the project and is granted a long-term right of access, usually 30 years. The major difference between DBFO and BOT is how the DBFO team is compensated. The DBFO partner is compensated through specified service payments during the life of the project. In highway projects in Great Britain, payments to the consortium include traffic-related payments based on "shadow tolls." **Shadow tolls** are payments made by the host government to the contractor on the basis of traffic flows at predetermined points along the roadway.

A main difference between DBFO and a BOT arrangement is that no actual tolls are collected from road users. In a BOT arrangement, the private sector recovers its costs through toll or fee collection, and there is no cost to the government for the construction of the project. With DBFO, the cost of the project in the form of annual payments is still ultimately paid by the host government. This means that there is a cost to the taxpayer with a DBFO arrangement. However, the cost of a DBFO project is less than that in the traditional method because efficiencies from private operation and construction reduce the overall cost of the project. A DBFO contract typically offers some protection to the private sector operator in the event that the government agency partner changes the conditions under which the road operates. This provides protection if other competing roads are upgraded during the contract period, thus reducing traffic flows.

In Great Britain, the DBFO project format has been used extensively. The types of projects that have been constructed in Great Britain using DBFO include roads, bridges, hospitals, and schools. The goals of the British Highway Agency have been to develop a private sector road operating industry that transfers significant risk from the government to the private sector, while minimizing project costs. Contractors on these highway projects are paid using shadow tolls. A study of the first four highway projects built in Great Britain using DBFO (National Audit Office, 1998) indicates the benefits of DBFO. Projects studied in the report ranged from the widening of 19 miles of expressway to the maintenance of 32 miles of highway. The report indicates that:

- The private sector takes significant financial risks on these projects, including the entire risk relating to design building and roadway operation.
- The core technical requirements of the project specified by the government owner should not be too detailed. The project should be defined to allow innovation and cost savings during construction by the builder.
- The project builder is selected through negotiation. There was a prequalification, and then four consortiums were selected to bid on each highway project. Bidders were then short listed. Negotiations were conducted between the government and the bidders. Each bidder submitted a schedule of shadow tolls as a basis of negotiation. The bidder that minimized the net present value of the shadow tolls was selected. This format of bidding requires the public sector bidder to estimate traffic flows over a 30-year period.
- Shadow tolls vary depending on traffic volumes. The shadow toll per vehicle is higher for low traffic volumes and lower for high traffic volumes. There is a cap on the volumes for which the government will pay shadow tolls. This removes the risk to government of traffic volumes being much greater than forecast, requiring a huge shadow toll payment.
- The advantage of the DBFO method is found principally in the freedom of design left to the concession company, the transfer of risks to the concession company, and the enhanced efficiency resulting from private management. Otherwise, the DBFO method would have no advantage over budgetary funding and would cost more because of more substantial financial expenses, stemming in particular from the required return on invested capital (Bousquet, 1999).

Examples of Public–Private Partnerships

In this section, examples of the use of BOT and DBOT are provided. These methods have been applied to infrastructure projects all over the world.

Examples of BOT Projects

Toll systems are in widespread use in eight European countries for roads and/or bridges and tunnels: Austria, Denmark, Spain, France, Greece, Italy, Norway, and Portugal. It has been found in the European countries that a BOT approach and toll systems are increasingly recognized as the most efficient means of replacing taxpayer money with user money. The State budget contribution to funding of the French national road system dropped from 56% to 22%, while toll revenue increased from 32% to 57% over the period 1973–1995.

Case Study: Dulles Greenway BOT

An example of a major BOT project in the United States is the Dulles Greenway, which was opened in 1995. The Dulles Greenway is a toll road that was built in Virginia using the BOT concept. The road extends 14 miles from Dulles International Airport to Leesburg, Virginia. The roadway connects to the existing Dulles Toll Road. The road is a four-lane limited-access highway within a 250-foot right of way. It is financed, built, and operated by a private consortium. The road required enabling legislation in the Virginia Assembly to allow the construction and operation of a toll road by a private company. A commission was set up to regulate applicants for toll roads, to supervise and control toll road operators, and to have responsibility for approving or revising toll rates charged by operators.

Autostrade International S.p.A. is a constructor, concessionaire, and operator of toll road networks in Italy. It is a general partner in the Greenway corporate entity and serves as the operator of the Greenway. Automated toll collection techniques are employed along with traditional manned toll collection booths. The total cost of the project was estimated at $326 million. Of the initial $68 million investment by the consortium partners, $22 million was for equity financing and the remaining $46 million provided access to various lines of credit that would serve as guarantees against project risks. A consortium of 10 lending institutions provided long-term financing of $202 million. The Greenway's primary benefit is that it allowed the roadway to be constructed in a period when no government funds were available for the project. Without the use of private sector funding, the project would not have been constructed (Levy, 1996).

Argentina has used BOT contracts to rehabilitate major sections of its road network. The goal of the program was reconstruction and maintenance of existing roads and a reduction of the public sector support required for highways. Bidding for the projects was competitive. In return for the right to collect tolls, the concessionaires were required to undertake a program of rehabilitation, maintenance, and capital improvements. There was some controversy with these projects because tolls were collected before rehabilitation work was completed. The proper oversight mechanisms were not in place from the central government agency. This illustrates the importance of developing proper relationships between the government and private sector to ensure project performance (Queiroz, 1999).

Examples of DBFO Projects

The Netherlands has implemented shadow toll projects that are similar to the British DBFO technique. The Netherlands has adopted the scheme for the construction of tunnels in the western part of the country. The objective is to construct a larger number of tunnels that would not be possible using budget sources alone. The Noord tunnel was the first for which private funding was adopted. The Dutch State Public Works Department allocated a lump sum of Fl 3.1 million for maintenance and operation over 30 years. This means that any increase in construction, maintenance, and operating costs incurred by the private consortium is borne by the state. The concession company provided the funds for construction and will continue as owner of the tunnel for 30 years, receiving remuneration for the investment according to the number of vehicles using the tunnel and the agreed tunnel fee (Bousquet, 1999).

The use of PPPs is not limited to transportation infrastructure projects. Other infrastructures such as schools and hospitals can be built using PPPs. The first hospital project constructed in Great Britain using a DBFO approach was the New Dartford and Gravesham Hospital. A private consortium was required to design, construct, and finance a new 400-bed hospital and then to maintain the hospital and provide support services for a period of up to 60 years. The National Health Service Trust (the British government agency responsible for health care in the United Kingdom) estimated that the discounted cost of the contract would be £177 million over the first 25 years that the hospital is in use, after which the National Health Service Trust could terminate the contract without penalty if it decides to close the hospital. The consortium is paid on the basis of its performance operating the hospital. It is expected that cost savings of 3%, or around £5 million, will be obtained, compared with an equivalent project using conventional procurement methods (Committee of Public Accounts, 2000).

Controversies Related to Public–Private Partnerships

The application of PPP is not without controversy. Primary fears are loss of control by the government agency, and fear that the PPP will be poorly managed and not deliver the promised performance. Additionally, there may be fears that the PPP will end up costing more than traditional procurement methods.

The privatization of maintenance for the London Underground has caused significant controversy. One of the prime reasons for the partnership is the desire to provide sustained investment in the underground, which had not been possible using tax revenues. The proposed PPPs will drive private investment of £13 million million over 15 years, with £8.7 million million spent on enhancements and £4.3 million million spent on maintenance (National Audit Office, 2000). PPPs are being formed that will designate three consortiums to maintain the London Underground. These infrastructure companies are planned to provide long-term investment planning, professional project management, and effective delivery of day-to-day maintenance for an annual payment. The trains and stations will still be run by the public sector. Payment is based on complex performance criteria. A primary fear of opponents of the scheme is that a divided management structure will ensue for the underground that will adversely affect operations and safety. It can be argued that the system will be parceled out to three private companies with little incentive to operate in a unified manner. It is feared that the government agency will lose control over the selection and management of major rehabilitation construction projects. Court challenges to the London Underground PPPs have been unsuccessful, and the projects have moved forward. This is an example of a complex joint venture PPP project that will require skillful coordination between the public underground and the private contractors performing the maintenance work.

Summary

PPPs are a method of project delivery that involves the private sector into the financing of infrastructure projects. The main impetus for their use is to allow projects to be constructed when no money is available in the public sector. Typically, a consortium of companies is formed to execute a PPP with construction contractors teamed with designers and financiers to execute the project. Two of the most important PPP methods are BOT and DBFO. In the BOT method, a private consortium finances and builds the project and then recoups its costs by collecting tolls from facility users. In the DBFO arrangement, a government agency pays the consortium directly over the life of the concession period but cost savings result because of the increased efficiency of the private sector. The use of PPPs will continue to grow as governments seek to meet the increasing demands for new infrastructure projects.

Key Terms

Build-operate-transfer (BOT)

Design-build-finance-operate (DBFO)

Public–private partnerships (PPPs)

Shadow tolls

Review Questions

1. What is a PPP and why is it becoming a more popular way of delivering infrastructure projects?
2. What are the major differences between the DBFO and BOT methods of PPPs?
3. How is a DBFO project similar to a more traditional design-build project?
4. What are the risks to PPP consortiums working on international projects?
5. Who are the typical team members in a PPP consortium?

Management Pro

MANAGEMENT PRO

Consider the infrastructure needs of your community. Are there any projects that could be done as a PPP? What are the current barriers to implementing this PPP? What are the potential benefits to your community? What different local agencies and companies would need to work together? Write a short report describing your findings.

References

Bousquet, F. 1999. Analysis of highway concessions in Europe: French study for the DERD/WERD. Washington, DC: World Bank. Available from http://www.worldbank.org/transport/roads/tr_docs/hway_conc.pdf (accessed August 21, 2002).

Committee of Public Accounts. 2000. The PFI contract for the New Dartford and Gravesham Hospital. House of Commons, Session 1999–2000, Twelfth Report. London: The Stationary Office.

Lawther, W.C. 2002. Contracting for the 21st century: A partnership model. Arlington, VA: The PricewaterhouseCoopers Endowment for the Business of Government.

Levy, S. 1996. *Build Operate Transfer*. New York: John Wiley.

Middleton, N. 2001. Public private partnerships—A natural successor to privatizations. London: PricewaterhouseCoopers.

National Audit Office. 1998. The private finance initiative: The first four design, build, finance and operate roads contracts. HC 476 Session 1997–98. London: The Stationary Office.

National Audit Office. 2000. The financial analysis for the London underground public private partnerships. HC 54 Session 2000–2001. Available at www.nao.gov.uk/publications/nao-reports/00–01/000154.pdf (accessed June 20, 2002).

National Council for Public–Private Partnerships. 2002. For the good of the people: Using public–private partnerships to meet America's essential needs. Available at http://ncpp.org/presskit/ncpppwhitepaper.pdf (accessed June 8, 2009).

Queiroz, C. 1999. Contractual procedures to involve the private sector in road maintenance and rehabilitation. Washington, DC: Transport Sector Familiarization Program, World Bank.

Savas, E. 2000. *Privatization and Public–Private Partnerships*. New York: Seven Bridges.

Smith, A.J. 1999. *Privatized Infrastructure*. London: Thomas Telford.

chapter 7

Construction Scheduling and Planning

Chapter Outline

Introduction
Basic Scheduling Methods
The Critical Path Method
CPM and the Computer
Summary
Key Terms
Review Questions
References

Introduction

Planning and **scheduling** are two related topics that are vital to the success of construction projects. Planning can be defined as the act of formulating a program for a course of action. Scheduling is the preparation of a timetable for the completion of the various stages of a complex project. Scheduling involves the coordination of many related tasks into a single time sequence. Implementation of a project requires careful planning and scheduling. A simple example is the building of a small garage. It is obvious that we must first build the foundation and the walls before the roof can be built. For complex projects, there are thousands of interrelationships between work activities. Contractors need to know answers to questions such as:

1. What activities need to be done first?
2. What activities depend on other activities?
3. What work tasks can occur simultaneously with other work tasks?

The need for good scheduling and planning of construction projects becomes more acute as construction projects become more complex. Construction projects can consist of tens of thousands of tasks. To prevent inefficiency and chaos, projects must be carefully planned and sequenced. Traditional methods of scheduling involved making bar charts of planned activities. However, with increasing project complexity, modern practice has turned to the use of computers and a scheduling algorithm called the **critical path method** (**CPM**) to schedule construction projects. The critical path method is a network-based technique that determines a connected chain(s) of activities through a project that can have no scheduling leeway if the project is to finish at the earliest possible time.

Basic Scheduling Methods

Bar charts (or Gantt charts) are one of the basic scheduling techniques. A bar chart is a graphical depiction of the amount of time required and the schedule of construction activities. Activities are shown as bars and are arranged vertically on the *y* axis of the chart. Time (in units such as days, weeks, or months) is plotted on the *x* axis. Figure 7.1 shows a bar chart.

The bar chart is a good way to show the schedule for simple projects or a small group of activities within a project. It also serves to provide an overview of major milestones on a project.

As projects become more complex, however, the bar chart becomes more difficult to understand because it may be many pages long. In addition, bar charts have some limitations in representing complex relationships between activities. A bar chart does not clearly show the constraints between activities. For example, Activity B in Figure 7.1 is

Figure 7.1

A simple bar chart

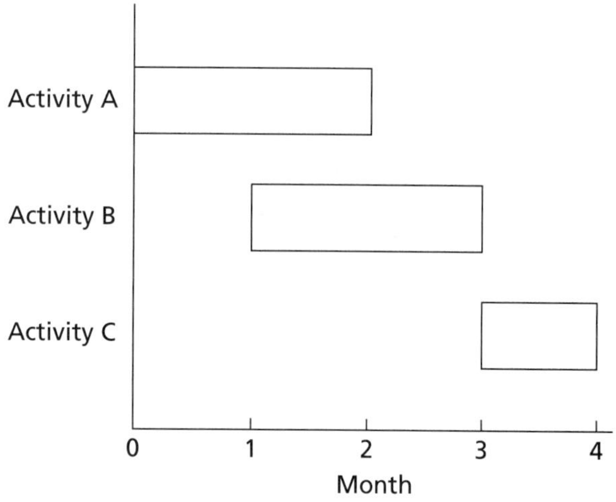

shown to start at the end of Month 1. Some questions immediately arise: Is Activity B related to Activity A, or is it independent? If Activity A is delayed, can Activity B still start at the end of Month 1? From the information provided in a basic bar chart, this is impossible to know. For this reason, more complex scheduling techniques, such as the critical path method, have evolved that consider the relationships between activities and the effects of delays on connected activities.

Another simple scheduling method is called a "**horse-blanket" schedule**. This method is typically used for the construction of highway or mass transit facilities where the work is spread out over a broad linear area. For example, the building of a subway line will occur over several miles of track along with the subway stations. Work along different segments of the line often progresses linearly with one type of work closely following another. An example is that after the concrete track bed is placed in the tunnel, cross ties can be installed, followed by the rails for the trains. Figure 7.2 shows a horse-blanket schedule for the construction of the subway system.

The Critical Path Method

The critical path method emerged in the late 1950s. It was developed as an application to manage the large projects that were beginning to emerge that were too complicated to be effectively managed with simple techniques like the bar chart. The first use of CPM was in 1956 by the E.I. DuPont de Nemours Company, which used a large mainframe computer for a chemical plant in Louisville, KY. Originally, CPM was only used by large, sophisticated organizations with access to mainframe computers. However, as computers have become smaller and faster, CPM is now available as a scheduling and planning tool for construction contractors of any size (Gould and Joyce, 2002).

Figure 7.2

A horse-blanket schedule for a subway construction project

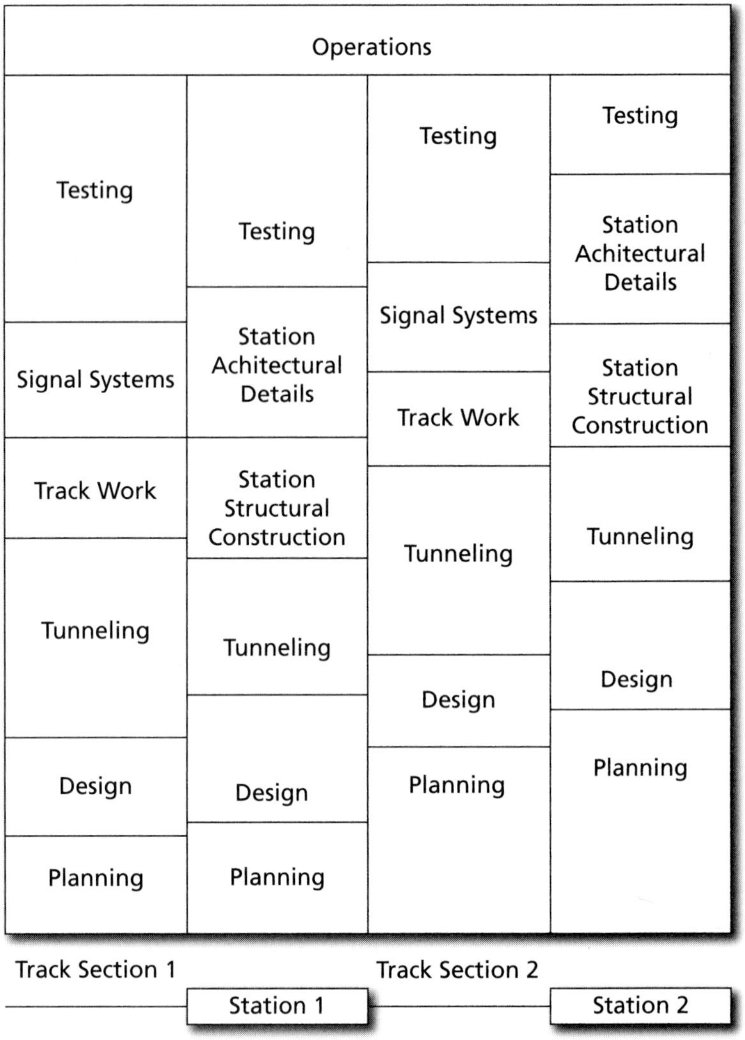

Example

The Importance of CPM in the Construction Industry

Using CPM to schedule construction projects is perceived as an important activity within the construction industry. A study of CPM software use by the ENR Top 400 Contractors has found that these large contractors generally consider the use of CPM as important to their companies' success. The study indicated that:

- Ninety-eight percent feel it is a valid management tool.
- Eighty percent believe it increases communication within the workforce.
- The Top 400 Contractors use CPM as a project-planning tool before construction and to periodically update schedules during construction.
- Contractors are increasingly using CPM during the estimating and bidding phases of a project to improve their understanding of the sequence of proposed project activities (Kelleher, 2004).

The Basics of CPM

CPM is a network-scheduling technique. The basic building blocks of a schedule, namely activities, are arranged as an interconnected network. An **activity** can be defined as anything that consumes time. A construction project must be decomposed into its individual activities. There are no hard and fast rules about the level of detail that is necessary: too few activities and the CPM network does not effectively represent the project and the interactions between important work tasks; if there are too many activities, the level of detail is so high that the model becomes difficult to work with.

After the activities are identified, the next step is to arrange them into a network. In this book, the activity on node type of network will be illustrated because most construction scheduling software employ this type of network. In construction scheduling, a network is drawn where each activity is represented by a node. Lines are drawn to connect the activities in the proper order. In a CPM network, no activity can commence until all its preceding activities have been completed. One of the first things to realize about a CPM schedule is that activities must be arranged in a logical manner. For example, a roof cannot be constructed until the walls that will hold it up are in place. Therefore, it can be seen that building a valid CPM model requires knowledge of the sequence of construction activities.

Developing a CPM Network: An Example

Imagine that you need to build a small shed. If you want to apply CPM scheduling to this small project, you must first understand the project constituents. The shed will have a concrete mat foundation and four wood-framed walls covered with an exterior sheathing. The shed will also have a shingled wood-framed roof. One wall will have a door. After the shed is constructed, the contractor will install some landscaping around the shed. Landscaping is typically installed at the end of projects so that it is not damaged. Consider the activities that will need to be performed, and then consider the logical order in which the activities must be arranged. It is also useful to consider the level of detail required to provide meaningful management information. If only two activities are used, as in Figure 7.3, if the second activity becomes delayed it would not be possible to ascertain the exact construction problem from the schedule. Logically, the level of detail of the schedule must be sufficient to provide the construction manager with meaningful insight into how the project is progressing.

Description of the required activity to build a small shed requires an understanding of what is physically possible. First, the foundation must be built so the walls can be made to bear on it. Then the walls must be framed. Only then is it possible to frame

Figure 7.3

A schedule with too little detail

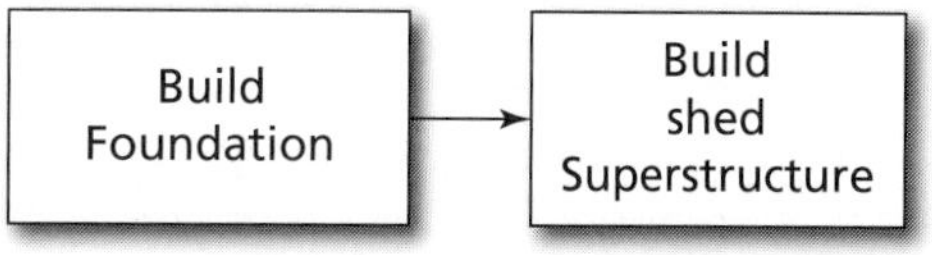

Table 7.1 Activity listing for shed construction project

ACTIVITY	IMMEDIATELY PRECEDING ACTIVITIES
Excavate foundation	—
Pour foundation	Excavate foundation
Cure concrete	Pour foundation
Frame Walls 1 and 2	Cure concrete
Frame Walls 3 and 4	Frame Walls 1 and 2
Exterior sheathing Walls 1 and 2	Frame Walls 1 and 2
Exterior sheathing Walls 3 and 4	Frame Walls 3 and 4
Frame roof	Frame Walls 3 and 4
Hang door	Exterior sheathing Walls 1 and 2
Shingle roof	Frame Roof
Landscape	Exterior sheathing Walls 3 and 4, shingle roof, hang door

the roof. Table 7.1 shows a listing of all the identified activities. Notice that there is a column to show each activity's predecessor labeled "Immediately preceding activities." This table provides enough information to draw the CPM diagram. It can be useful to construct a table like this before drawing the CPM diagram.

Other activities can follow after the superstructure of the building has been constructed. The roof can be shingled and the wood sheathing can be placed on the exterior. At some point after the wooden sheathing has been placed, the door can be hung. Figure 7.4 shows the CPM diagram that can be constructed for building the shed. Notice that the network allows activities that are not dependent on one another to

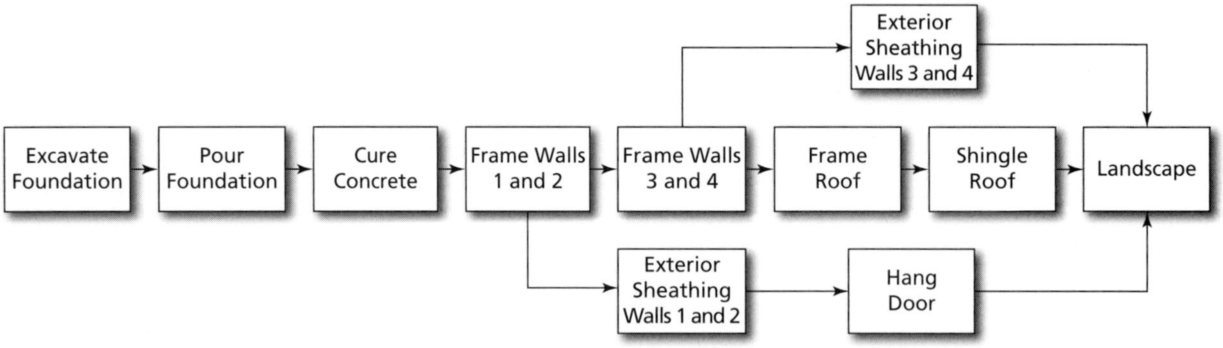

Figure 7.4 A CPM diagram for the shed project

proceed in parallel. It should also be noted that the schedule could be more detailed or some of the project work tasks could be broken down in different ways depending on the construction constraints and the construction methods used.

Determining Activity Durations

Use of CPM requires the user to input a deterministic value for the activity duration. The time unit used can be of any type (minutes, hours, days, etc.). The use of time units must be consistent for all activities in the network. In the construction industry, activity durations are typically given in days. Construction managers often assign activity durations on the basis of their experience. They may also use company records from past projects that describe how long an activity took. The unique nature of many construction projects and their complexity can sometimes make it a challenging task to assign activity durations. During construction, as more information becomes available about crew productivities that will affect durations, it is important to update the schedule on a regular basis.

CPM Calculations

The calculation of the critical path and the project duration is relatively simple, requiring only addition and subtraction. Activities in a critical path network have an early start and an early finish time. They also have a late start and a late finish.

Forward Pass

The first step in the calculation is to perform a **forward pass**. In this step, the early start and early finish of each activity are calculated. The **early start** (**ES**) is the earliest time an activity may start. The **early finish** (**EF**) is the earliest point at which an activity can be completed. Initial activities in the network are assumed to have an early start time of zero. All other nodes in the network are calculated using the following formula:

$$ES_i = \text{maximum EF of all preceding activities}$$

The maximum value of all the preceding activities' early finishes is selected as the early start. Then the early finish is obtained by adding the early start and the activity duration:

$$EF_i = ES_i + DUR_i$$

Moving forward through the network (from left to right) leads to calculation of the early start and early finish for each activity. The early finish of the final activity gives the project duration.

Backward Pass

The second stage of the CPM calculation is to move backward through the network and determine the late finish and late start times for each activity. The **late finish** (LF) for the last activity in the network is assumed to be equal to the early finish calculated

Table 7.2 Example network activities and durations

ACTIVITY	IMMEDIATELY PRECEDING ACTIVITY	DURATION (DAYS)
A	—	3
B	A	6
C	A	5
D	B	7
E	C	6
F	D, E	4

in the forward pass. If there are multiple closing activities, the greatest early finish is used. All other nodes are calculated using:

$$LF_i = \text{Minimum LS of all following activities}$$

Then the **late start** (LS) can be calculated as:

$$LS_i = LF_i - DUR_i$$

Example Network Calculations

Table 7.2 provides the information necessary to draw the network shown in Figure 7.5. Figure 7.6 illustrates the meaning of the numbers shown on each node. At this point, the forward pass can commence. The results of the forward pass are shown in Figure 7.7. Notice how the forward pass equations are applied. The early start of Activity A is at time zero. The early finish of Activity A is the duration of Activity A plus its duration. As defined earlier, the early start of an activity is the earliest possible time it can commence and the early finish is the earliest possible time the activity can be completed. Therefore, it is logical that the early start of the first activity is at

Figure 7.5

Example CPM network

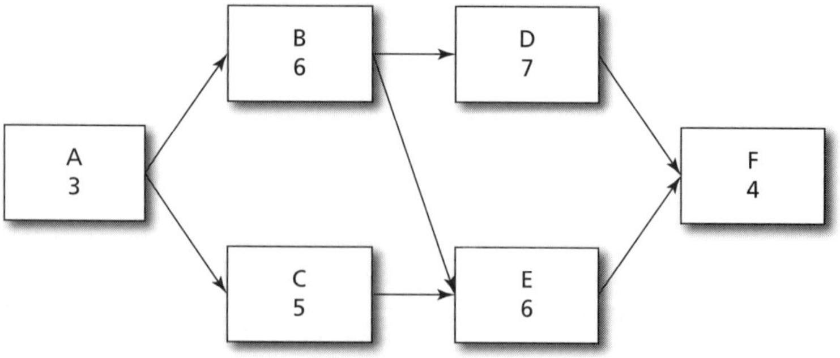

Figure 7.6

Explanation of numbers shown on nodes

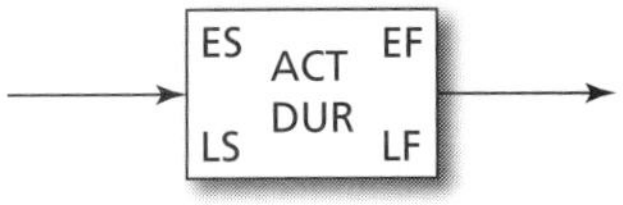

7

Figure 7.7

Example network after forward pass is completed

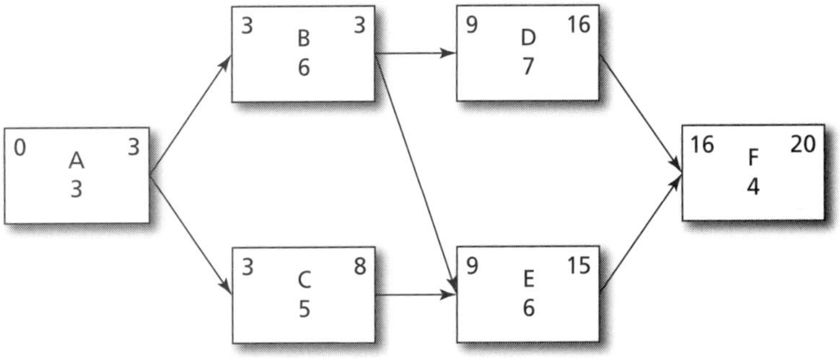

time zero and its early finish is its duration. The early start of Activities B and C are the early finish of activity A. The early finish of B and C are equal to their early start plus the activity duration. So for B, the early finish is 9, and for C, the early finish is 8. Activity E provides a good example of how to apply the forward pass calculation to activities preceded by multiple nodes. For the early start of E, we must select the maximum of the early finish of Activities B or C, because they are the two nodes that precede E in the network. Therefore, the early start of E must be on Day 9, the early finish of Activity B, which is 1 day more than the early finish of activity C. Activity F is preceded by Nodes D and E. Again we take the maximum finish time of the preceding activities, D and E, as the early start of F. The early finish of F gives the project duration, 20 days.

The backward pass commences by making the late finish of the final activity F equal to the early finish. This can be done because the last activity must obviously be complete as early as possible because it is on the critical path and will not have any scheduling leeway. The late start of F equals the late finish minus the activity duration. Moving backward through the network from right to left, the late finish of Activities D and E is equal to the late start of Node F. Activity B is followed in the network by two activities, D and E. As we move backward, we compute the late start times of D and E as 9 and 10 days respectively. Here, the main differences between the forward and backward passes are highlighted. As the late finish of B, we select 9, which is the *minimum* of the late starts of the preceding activities. The late start of B is Day 3, which is the late start minus the activity duration (9 – 6). The completed network after the backward pass is shown in Figure 7.8. More information can be extracted from this network but the idea of scheduling leeway or float must be discussed first.

Figure 7.8

Example network after backward pass is completed

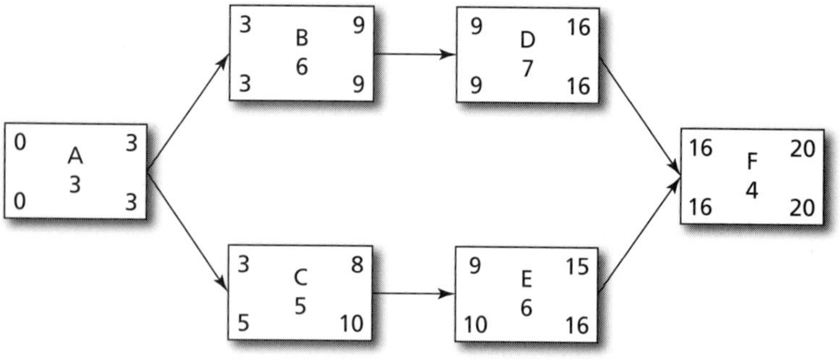

CPM Float

The goal in any CPM calculation is to identify the minimum path through the network to identify the amount of time the project will take. Some activities will be "critical." **Critical activities** cannot be delayed or else the duration of the project will be longer. Critical activities reside on unbroken chains through a CPM network. Therefore, a critical path or paths can be identified for a project using the CPM technique.

Those activities that are not on a critical path will have scheduling **leeway**, meaning that their start times can be adjusted and within limits, this will not affect the completion date of the project. In the construction industry, this scheduling leeway is commonly called **float.** There are various types of floats that can be calculated. In the construction industry, **total float** is widely used. This is because the scheduling software that is most commonly used in the industry provides total float as an integral part of their output.

Total float is defined as the amount of time for which an activity's start can be delayed without affecting the project duration. Total float is commonly shared along connected chains of noncritical activities. Therefore, if an early activity is used, it may alter the starting times of following activities and could make activities that originally had float to become critical later in a project. This underlines the importance of regularly updating schedules for any construction project. The total float for an activity can be calculated as:

$$TF_i = LF_i - EF_i = LS_i - ES_i$$

It is important to understand that a critical activity will always have zero float and that only activities with scheduling leeway will have float. Therefore, the total float calculation can serve two purposes. First, it can be used to identify all critical activities and determine the critical path(s) in a network. Second, it identifies the magnitude of total float for those activities that are not on the critical path.

A second type of float that can prove useful to construction managers is **free float,** which is the amount of time for which an activity can be delayed without affecting the starting time of the following activities or the completion time of the project. If an activity has zero total float, it cannot have any free float. The formula for calculating free float is:

$$FF_i = \text{Minimum early start of all following activities} - EF_i$$

Table 7.3 Float calculations for example network

ACTIVITY	TOTAL FLOAT	FREE FLOAT
A	3 – 3 = 0	3 – 3 = 0
B	9 – 9 = 0	9 – 9 = 0
C	10 – 8 = 2	9 – 8 = 1
D	16 – 16 = 0	16 – 16 = 0
E	16 – 14 = 2	16 – 15 = 1
F	20 – 20 = 0	0

Calculating Floats for the Network Example

Using the formulae for the total float and free float, it is now possible to return to the schedule shown in Figure 7.7 and identify the critical activities and calculate the floats of the noncritical activities. Table 7.3 shows the float calculations.

Examination of Table 7.3 shows that the critical path through the network is the activities A → B → D → F because these activities have no scheduling float. Activities C and E have both float and total float. Activity C has 2 days of total float and 1 day of free float. If Activity C is started on Day 4, it will not affect the early start of Activity E. So, if Activity C is started a day late, it will not delay the commencement of E or the completion of the project. If Activity C is not started until Day 5 of the project, the 2 days of total float will be used up and Activity E will have to start at its late start time. This is an example of total float where the start time of the following activity is affected but the total project duration remains the same.

CPM and the Computer

The use of the CPM is closely related to the development of the computer. For simple projects, it is relatively easy to draw and maintain a CPM schedule by hand. However, most construction projects have hundreds or thousands of activities that must be tracked; so the use of the computer to calculate and maintain becomes a necessity. Those readers who will pursue a career in construction will find that an understanding of how to use CPM software is often a necessity.

Construction companies typically use two competing software packages. They are Primavera and Microsoft Project. CPM software such as Primavera and Project offer many powerful features beyond the basic CPM calculations.

Basic Functions of CPM Software

The core function of any CPM software is to calculate the project duration, identify the critical activities, and identify the total float for noncritical activities. Importantly, with the continuing development of computer graphics and printing capabilities, CPM programs now offer different types of output that provides a clear representation of the schedule that is easily used by various types of users. Most scheduling programs now provide the output of the scheduling information in the form of color-coded bar charts, time-scaled network diagrams, and tabular reports. The software allows information to be tailored to various types of users, such as reports for top management, and simple bar charts that are understandable to users at the construction job site.

The basic features of CPM software are summarized:

- Input forms are available to name activities, code activities, input durations, and input the relationships between activities.
- The CPM software calculates the project duration and the completion date of the project.
- Critical activities are determined and flagged. Typically, the program output clearly identifies critical activities for the software user.
- Noncritical activities are identified and their floats calculated. Programs have controls that allow users to select when noncritical activities start (such as early start or late start).
- Calendar management is available, which takes into account work stoppages on weekends and holidays. Multiple calendars are maintained.
- High-quality reports and charts are generated to disseminate schedule information, including bar charts and other types of reports. Programs typically have the capability to print bar charts in color and to print high-quality versions of the project network.
- The software allows for users to update activity details and recalculate schedule details. CPM programs typically allow different "versions" of the schedule to be maintained so that the original project plan can be compared with the current project schedule.

Some examples of how data are input to a scheduling program are shown in Figures 7.9 and 7.10. Figure 7.9 shows how data can be input using the Gantt chart view of Microsoft Project. Users input an activity name and duration. Highlighted activities can be linked together using the link icon on the toolbar. The project Gantt chart is displayed on the right of the screen and updates as changes are made to the scheduling activities. Figure 7.10 shows the Network Diagram view in Microsoft Project. In this view also, new activities can be created, and activities can be linked by dragging an arrow between two activities. This provides a rapid way of inputting the relationships between activities.

Figures 7.11 and 7.12 show a bar chart and a CPM diagram produced by the Primavera scheduling software. Figure 7.11 shows a portion of a bar chart produced for an actual process plant retrofit. In the figure, the wide bars indicate the actual schedule. The thin

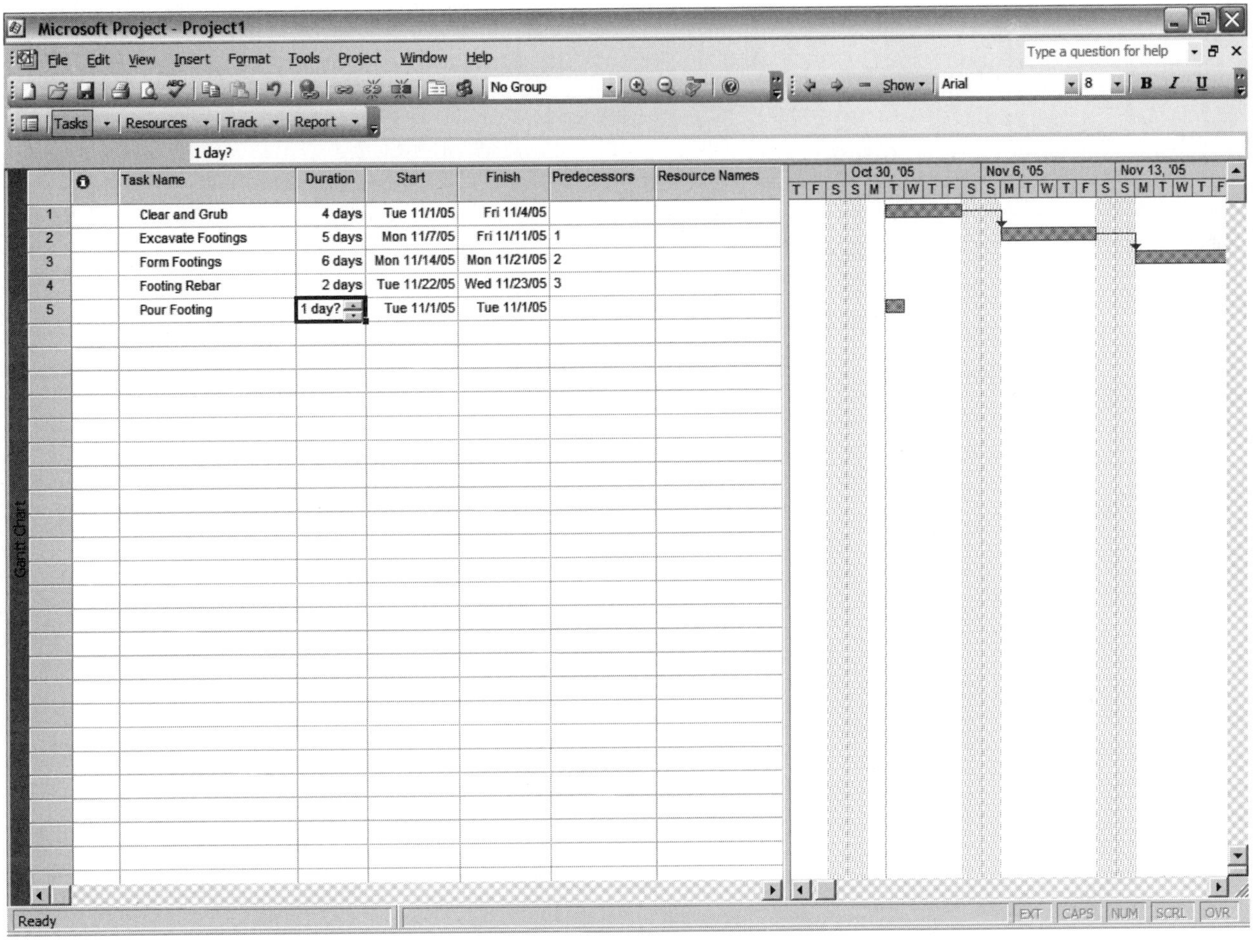

Figure 7.9 Input of scheduling data using Microsoft Project: Bar chart view

black bars show the original project plan. A careful examination of the bar chart shows that the sequence of activities has been modified during construction. Using Primavera, the same information in the bar chart can be viewed in network form.

Figure 7.12 shows a CPM network screen display produced by the Primavera software. This is actually a network drawing showing the interrelationships among activities, duration, and total float. Those activities with a slash across them are those that have been completed. These figures illustrate the capability of scheduling software to communicate schedule information.

Advanced CPM Software Functions

Many CPM computer programs offer sophisticated methods for enhancing the management of project schedule and costs. Several are discussed in the following text.

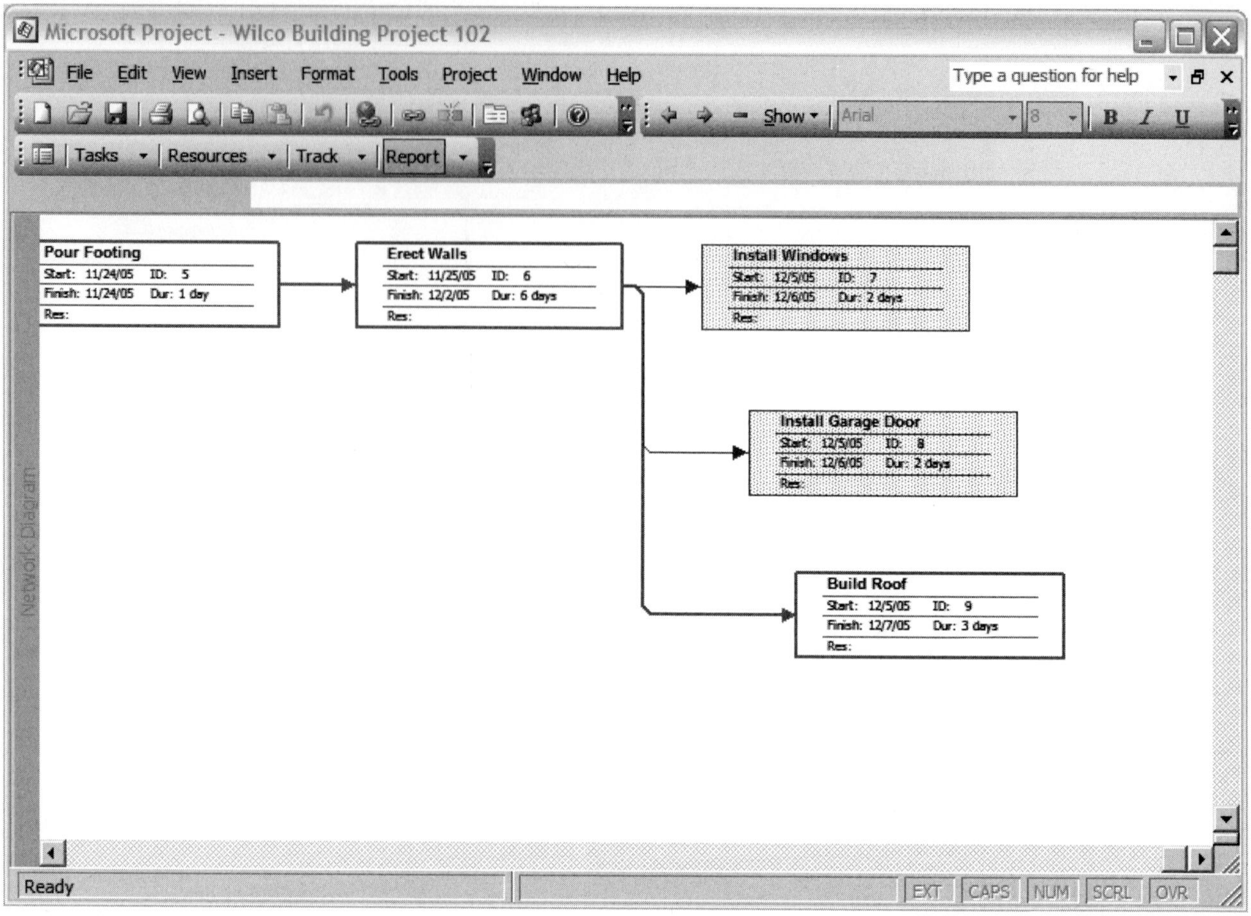

Figure 7.10 Network diagram in Microsoft Project

CPM and Project Cost Control

CPM scheduling software has the capability to track costs for each scheduling activity. Construction companies often use programs such as Primavera to provide integrated control of the schedule and project budget. In other words, it is possible to determine if a scheduling activity is late and to determine if an activity is over budget from the CPM software. The following points highlight aspects of cost handled by CPM software:

- "Cost loaded" schedules are available where each activity has a budgeted cost assigned to it and actual costs are input to the CPM software during construction.
- Construction budgeted versus actual costs can be compared for scheduling activities.
- CPM software is typically able to generate S-curves of project expenditures.
- Project resources such as labor and equipment can be associated with each scheduling activity and their costs tracked.

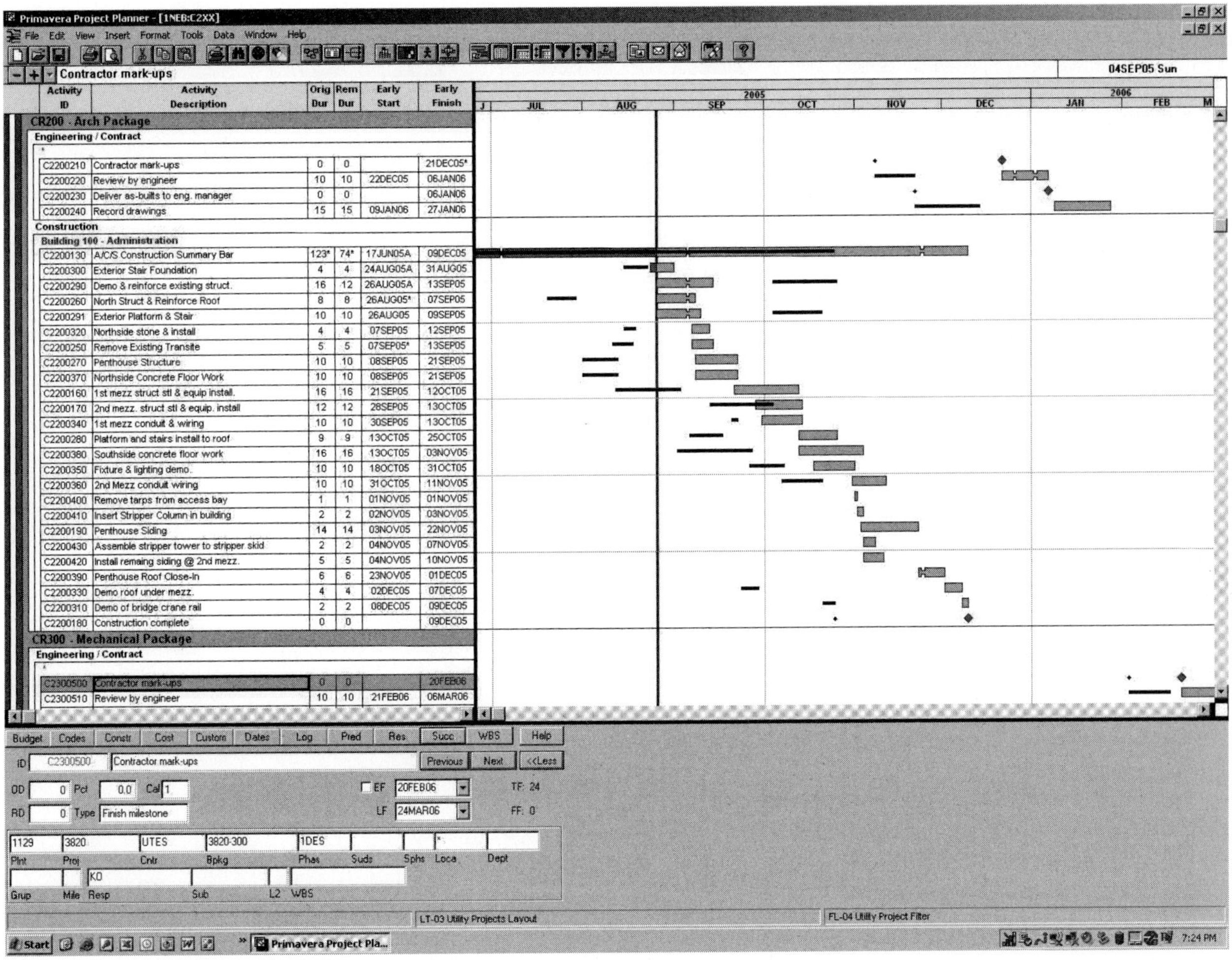

Figure 7.11 Bar chart for an actual process plant retrofit using Primavera

Progress, Schedule, and Cost Control

Imagine you are the president of an international construction company and you are given a status report about a project. You are told that a project has used 15 of its allocated 30 days. You are also told that about $150,000 of the budgeted $400,000 has been spent. Is this enough information to determine if this project activity is proceeding smoothly? Possibly, you could draw the conclusion that because 15 days have elapsed (one-half of the scheduled project duration) and $150,000 has been spent, the project is on schedule and under budget. However, more information is needed to draw conclusions about the actual progress that has been made. Fifteen days may have elapsed since the start of the activity but we do not know if all those days have been productive. The president of the construction company knows that $150,000 has been expended but he or she does not know if this represents the actual value added to the

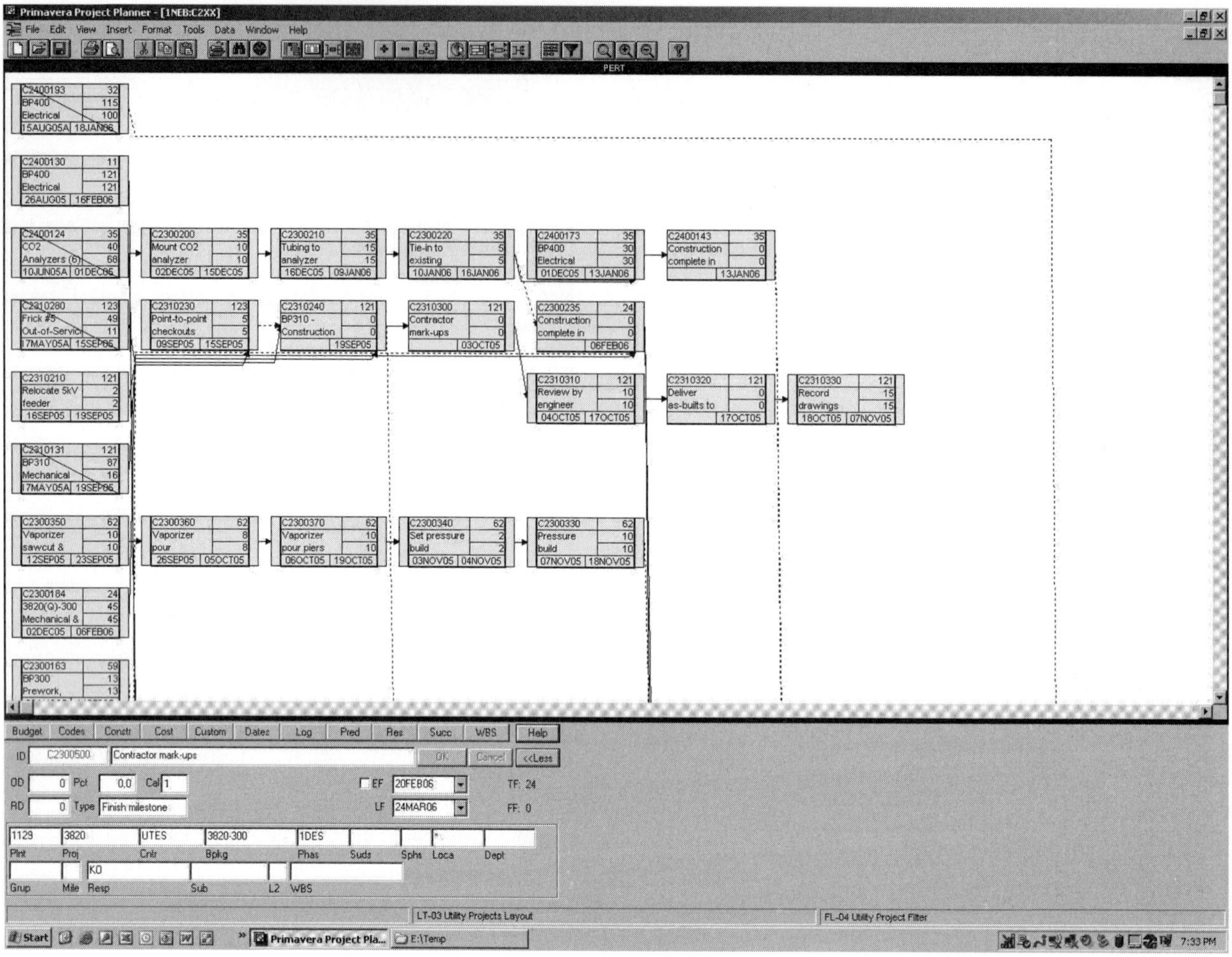

Figure 7.12 Process plant project CPM network

constructed facility or if money has been wasted because of poor management in the field. The cost control features of advanced CPM programs allow for the calculation of earned value (EV) from data input from the field. With some additional input from the field about the actual value of the construction work performed, CPM software can perform the **earned value** calculations automatically to provide managers a better understanding of the project status.

To use EV as a clearer measure of progress on a construction project, several values must be calculated to provide a clearer picture of project progress. They include:

- **Budgeted cost for work scheduled** (**BCWS**). The original amount budgeted for the activity at the start of the project.
- **Budgeted cost of work performed** (**BCWP**). The actual earned value of the work that has been performed to date. It is the budgeted cost for work that has actually been performed.

- **Actual cost of work performed (ACWP).** A measure that integrates schedule progress and costs for an activity because it includes monitoring of time (work performed) and cost records (Hinze, 2008). It gives the actual costs for an activity to date.

This method requires experienced personnel who can determine the actual earned value of construction activities. Additionally, the earned value technique requires additional information to be input for each activity in a network and requires commitment from the construction contractor to maintain up-to-date data in the scheduling program.

Let us return to the construction company president's dilemma. He or she needs to know the BCWS and BCWP to make conclusions about progress. The president already knows that ACWP is equal to \$150,000. He or she makes a phone call to the field and learns from the project manager that the BCWS was \$200,000 and that the BCWP is \$100,000. Some simple calculations can provide greater meaning to these numbers. First, we can calculate the schedule variance for the activity. The schedule variance is:

$$SV = BCWP - BCWS$$

For this example the SV is:

$$SV\ (example) = \$100{,}000 - \$200{,}000 = -\$100{,}000$$

The negative number indicates that the project is behind schedule. A positive number would indicate that it is ahead of schedule. Similarly a cost variance (CV) can be calculated:

$$CV = BCWP - ACWP = \$100{,}000 - \$150{,}000 = -\$50{,}000$$

This result indicates that the project is over budget as well. A negative number indicates that actual costs are greater than budgeted costs. Two performance indexes can also be calculated. They are the schedule performance indicator (SPI) and the cost performance indicator (CPI). For this example the SPI is:

$$SPI = BCWP/BCWS = \$100{,}000/\$200{,}000 = 0.5$$

The CPI can be calculated as:

$$CPI = BCWP/ACWP = \$100{,}000/\$150{,}000 = 0.67$$

These numbers indicate that the project is both behind schedule and over budget. A ratio of 1 for the SPI would indicate that the project is on schedule. A ratio of 1 for the CPI would indicate that the project is meeting its budgeted target. Ratio values greater than 1 would indicate that the project is being built ahead of schedule and for less than the budgeted amount, respectively. Advanced scheduling software allow for input of cost data for each activity, including the budgeted cost, actual costs, and the amount of progress to automatically make these earned value calculations and to provide them to project decision makers.

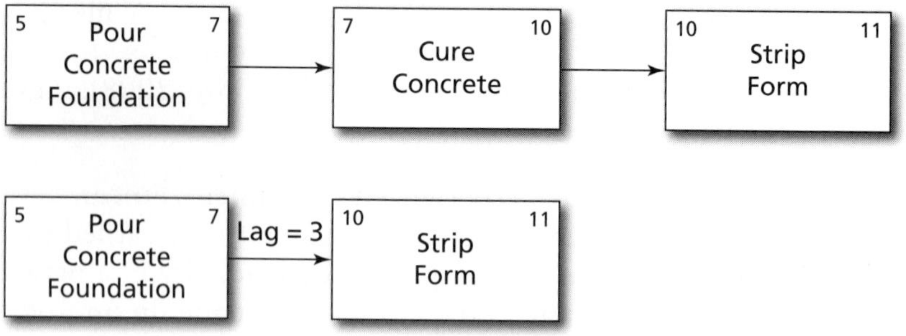

Figure 7.13 Use of lags to reduce activities

Advanced Activity Relationships

The examples in the chapter so far have used only a finish-to-start relationship between activities, that is, only networks where a following activity starts when its preceding activity is complete have been studied. There are several other scheduling relationships between activities that are possible, some of which are described here:

- **Lags** are used where the finish-to-start relationship is maintained but the following activity is delayed for some number of time periods specified by the scheduler. Figure 7.13 shows the classic example where rather than using an activity that takes no resources to represent the time for concrete curing, a lag is used to delay strip forms.
- A **finish-to-finish** relationship where two activities are designated to finish on the same day. This is shown in Figure 7.14 where the scheduler has decided that road stripping and install signs must end at the same time.
- A **start-to-start** relationship is where two activities must start at the same time. In Figure 7.15, a start-to-start relationship is used to reduce the number of activities. This represents the idea that it is possible to start paving the driveway before the grading is complete.

Computer software allow determination of these different relationships between activities. However, caution must be used in using these tools because they can make

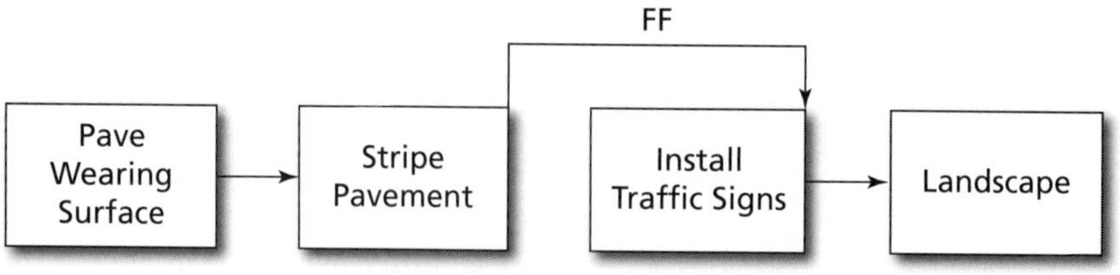

Figure 7.14 Example of finish-to-finish activities

Figure 7.15

Example of start-to-start activities

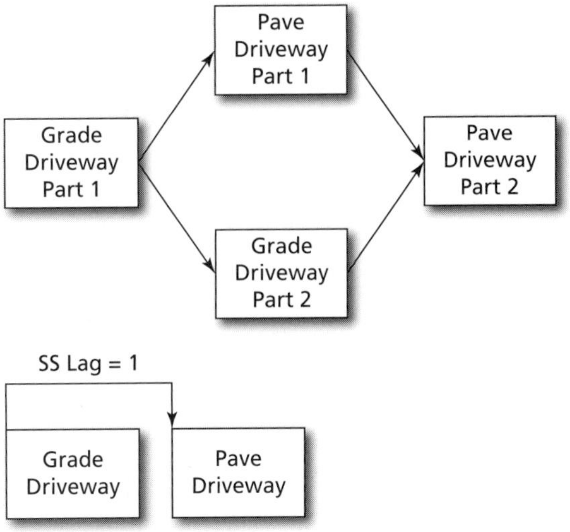

the schedule more difficult to understand and can make float calculations produced by the CPM software difficult to interpret (Oberlender, 2000). Therefore, it is recommended that they be used sparingly and only be used when they are needed to represent complex relationships between activities when the straightforward finish-to-start relationship cannot represent them.

Case Study: Applications of CPM in the Construction Industry

The following article excerpt describes a company that uses CPM to produce construction schedules for contractors. This article titled "New Firm Aims to Help Contractors Deliver on Time: Increasingly Complex Projects Are a Growing Scheduling Challenge" was published in the *ENR* magazine on November 13, 2006. Tom Nicholson wrote the article.

> In a market where owners increasingly invest in fast-tracked projects of unprecedented scope and complexity, many contractors' project-scheduling resources are being pushed to the limit. But for one national firm, today's tough scheduling demands represent an emerging market to be tapped.
>
> Many factors go into successfully delivering a project, "but at the end of the day it's the schedule that matters," says Dave Ambrose, president of Turner Construction Co.'s newly formed subsidiary, Quality Planning Solutions Inc., which claims to be one of the largest project scheduling groups in the nation.
>
> Reston, Va.-based QPS was launched last May as Turner executives eyed the burgeoning market for schedulers that the industry boom has groomed. "We wanted to add a service to distinguish Turner from other contractors," says Ambrose. "By becoming a separate entity, it allows us to continue scheduling for Turner, but be able to go outside and offer the service to others."

QPS offers "advanced schedule development, change management expertise, schedule recovery planning and change recovery analysis that is required on many of today's complex projects," says William M. Brennan, executive vice president of New York City-based Turner.

Seven months after hitting the market, QPS so far has handled construction project scheduling for about eight outside clients in addition to scheduling about 50 Turner projects, says Ambrose. "We've set up critical-path-method schedules, and now we are doing monthly updates throughout each job," he says.

Dulles, Va.-based contractor M.C. Dean Inc. is using QPS's services on two projects—a $3.5-million security upgrade at Orlando International Airport and a $13-million security upgrade for the Metropolitan Atlanta Rapid Transit Authority (MARTA). "We've had to bring in schedulers because the projects we do have gotten much bigger than what we were typically used to," says Greg Kraning, Dean program manager. "Four years ago, we did projects in the half-million-dollar range. Now, we have numerous projects in the $13-million to $15-million range. These kinds of projects require a totally different skill set to schedule and manage."

Delays on the MARTA project have set the schedule back by about three months, a problem that Kraning says has been tempered by the outside scheduling. "There were many changes on this job," says Kraning. "There are things we could have done better, but once we started calculating out the delays, it got very confusing and QPS has been providing support."

Ambrose says the firm has 16 schedulers on staff, who can tap into Turner's broad experience to reference how to best schedule current projects and react to changes. "We have an archive of past project schedules and we look at general time frames of similar jobs and try to identify any differences," he says. "It helps owners and contractors bring in a second set of eyes, someone who can say 'This is what we've done in the past.'"

Rising Demand

QPS can "work with any type of project but it's best when we can get in very early," says Ambrose. "Design-build, fast-track projects or renovations and additions, for example, are very complex and can be very challenging to schedule, and the more complex the job, the more we can be of help." He adds, "Any project over $10 million is when you start needing a scheduler."

Owners' demands helped the firm determine where scheduling services were needed most in the market. When owners are selecting project teams, "they will say they want someone on staff with at least five years of scheduling experience and can work with Primavera scheduling software. Often they have to go outside for that," says Ambrose.

Headhunters have seen demand for scheduling skills shoot skyward in recent years. "In the last three years, the need for scheduling talent has become an everyday requirement. It's not a secondary function like it once was," says Frank Bruckner, executive vice president at Asheville, N.C.-based executive search firm Kimmel & Associates. Bruckner says contractors of all sizes now regularly seek project managers with strong scheduling skills.

The rise in demand is due partly to new scheduling technology available, but also because owners want to build better, faster and cheaper, Bruckner says. "Having scheduling skills on staff is no longer a luxury, it's an absolute necessity," he says. Today's hyperactive market has prompted some firms to create a new staff position called "progress manager," who rides herd on a project from beginning to end, Bruckner explains.

Jim Hovey knows a lot about what makes a schedule work, or not. As a project controls manager for Pasadena, Calif.-based Parsons Corp., which is managing construction projects at Baltimore/Washington International Airport, it's Hovey's job to review construction schedules for the 25 to 30 projects that take place at the massive airport each year. "Scheduling is fairly new," Hovey says. "It didn't become a real function until the invention of Primavera a few years ago." In working with contractors on the airport's projects, "most don't have a scheduler, but we recommend they do," Hovey says.

Hovey is working with QPS schedulers on two jobs that P. Flanigan & Sons Inc., Baltimore, is performing at BWI—a $28-million concrete apron project and a $13-million runway paving job. "Many times when we recommend a contractor get a scheduler, they end up going to the yellow pages and often we aren't satisfied with the results," He says. Reacting to scheduling snags is key. "It's important to realize that every project has changes, what matters is how they are handled," he says.

Clients invoice Turner for scheduling services provided by QPS, based on a lump-sum agreement. While costs for services will vary by geographic location, included are an initial setup and monthly updates. "If we hit a scheduling snag, we also will give the client a time-impact analysis," says Ambrose.

QPS uses multiple software such as Primavera P3.1, Sure Trak and Prolog for scheduling and submittal controls. Critical path schedules are created in P3.1, and Sure Track is used in the field, says Ambrose.

Turner's venture into project scheduling is a sign of the times, and it may be a glimpse of the future. "There is more and more need to speed up projects," says Chris Hendrickson, an engineering professor at University of Pittsburgh, who authored the widely used book *Project Management for Construction*. "As you speed up a schedule, you reduce built-in buffers, that's why project scheduling will continue to be critical," he says.

Summary

This chapter has shown how construction scheduling has evolved from the use of simple bar charts to the widespread use of the critical path method using computers. CPM scheduling has been greatly aided by the evolution of computing technology and the development of sophisticated computer programs capable of scheduling complex projects with thousands of activities. Beyond the basic scheduling functions of calculating the project duration and floats, advanced users of scheduling software can combine cost control and scheduling and perform earned value analyses. CPM scheduling is a standard technique in the construction industry and its use is becoming more widespread as projects increase in complexity.

Key Terms

Activity
Bar chart
Critical activities
Critical path method (CPM)
Early finish (EF)
Early start (ES)
Earned value (EV)
Finish-to-finish
Float
Forward pass
Free float
"Horse-blanket" schedule
Lags
Late finish (LF)
Late start (LS)
Leeway
Planning
Scheduling
Start-to-start
Total float

Review Questions

1. Two noncritical activities are linked in a start-to-start relationship, as shown in Figure for Problem 1. Determine the late start of B and the late start and early finish of A.

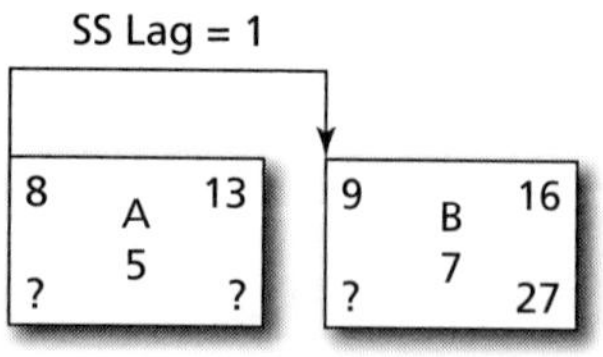

Figure for Problem 1

2. Three activities are shown in Figure for Problem 2. Activities A and B are linked by a finish-to-finish relationship with a 2-day lag. What are the early start and early finish times of Activities B and C?

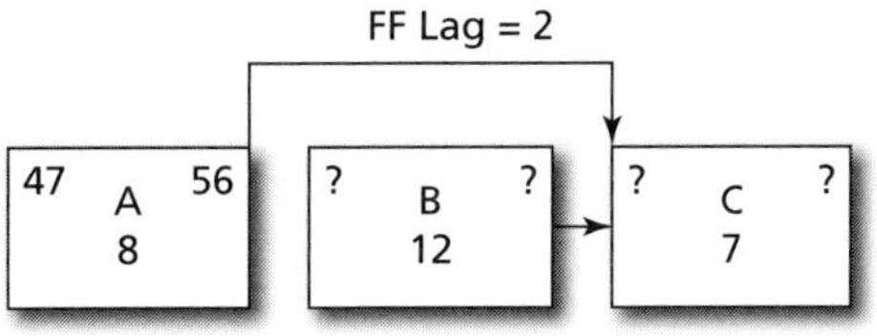

Figure for Problem 2

3. Given the following table, draw the CPM network; calculate the project duration and total and free floats for each activity. Identify the critical path(s).

Homework: Problem 3 network

ACTIVITY	IMMEDIATELY PRECEDING ACTIVITY	DURATION (DAYS)
A	—	3
B	A	4
C	A, B	3
D	C	2
E	B, D	4

4. Assume that the Problem 3 project is in progress. It is the end of Day 9 and you have been provided with cost data from the field in the following table. Calculate the CV, SV, CPI, and SPI for the project. Is the project ahead of schedule? Is the project below budget?

Homework: Problem 4 data

ACTIVITY	% COMPLETE	BUDGETED COST	BCWS	BCWP	ACWP
A	100	1,000	1,000	1,000	1,000
B	100	2,000	2,000	2,000	2,500
C	33	1,500	1,000	500	750

5. Given the following table, draw the CPM network; calculate the project duration and total and free floats for each activity. Identify the critical path(s).

Homework: Problem 4 network

ACTIVITY	IMMEDIATELY PRECEDING ACTIVITY	DURATION (DAYS)
A	—	3
B	A	4
C	B	3
D	B	2
E	B	4
F	D, E	4
G	E	6
H	C, D	5
I	F, G, H	4

6. Given the network shown in Figure for Problem 5, calculate the project duration, identify the critical path, and calculate the total path of each activity. Hint: LS_g and LS_h = max (EF_g or ES_h).

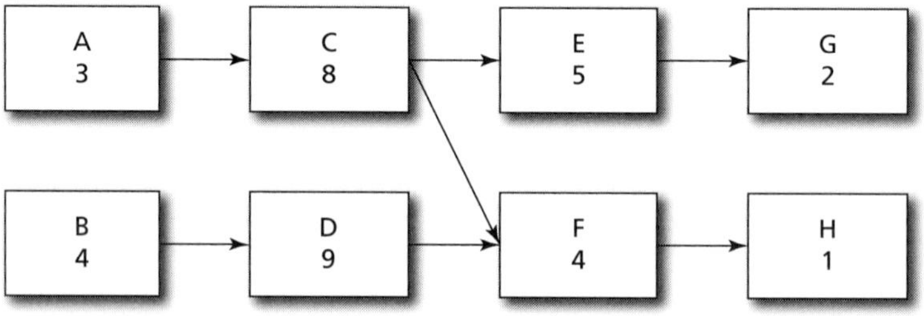

Figure for Problem 5

Management Pro

For the bridge and traffic circle shown in Figures 2.8 and 2.9, identify the major scheduling activities. Draw a CPM network for the project. Compare your network diagram with those of your classmates. How do your networks differ? Do you have more or fewer activities? Why?

References

Gould, Frederick E. and Joyce, Nancy E. 2002. *Construction Project Management.* Upper Saddle River, NJ: Prentice-Hall.

Hinze, Jimmie. 2007. *Construction Planning and Scheduling*, 3rd ed. Upper Saddle River, NJ: Prentice-Hall.

Kelleher, Andrew H. 2004. An investigation of the expanding role of the critical path method by ENR's top 400 contractors. Master's Thesis, Virginia Polytechnic Institute and State University.

Oberlender, Gerald. 2000. *Project Management for Engineering and Construction.* New York: McGraw-Hill.

chapter 8

An Introduction to Construction Cost Estimating

Chapter Outline

Introduction

An **estimate** is an approximate calculation of the degree of work, that is, cost of a construction project. Estimating is important in the construction industry. Construction contractors may depend on estimating success to win their project. Estimating can be difficult and initial cost estimates are exceeded frequently. Flyvbjerg et al. (2003) studied how very large projects are often constructed for sums greatly above original cost estimates. In that study, data from 258 transportation infrastructure projects, including bridges, tunnels, highways, freeways, and high-speed rail, urban rail, and conventional rail projects, were analyzed. The study found that 9 out of 10 transportation infrastructure projects were underestimated, resulting in cost overruns. The average cost increase on the projects was 28%. One possibility is that early stage estimates from enthusiastic promoters of a project may be low to gain project approval and funding. Then, as the project unfolds, the initial estimates turn out to be much too low. In another similar study (Love, 2002), cost growth on 161 Australian building construction projects procured using various contract types was investigated, and average cost growth on the projects was found to be 12.6%.

Factors Influencing the Cost of a Construction Project

Many factors influence the cost of a construction project. Two interrelated factors are the size and complexity of a project. The many factors that contribute to project complexity include:

- **Size.** A project may be so large that it becomes difficult to estimate. In a very large project, it may be difficult to manage and control cost. Many large projects bring materials, labor, and equipment together in a scope that is not normally seen. Often, a megaproject can cause localized labor and material shortages, which have inflationary effects that could not have been foreseen at the time of the estimate.
- **Complexity.** Complex projects tend to have lower productivity and cost more to build. Some examples of the complex buildings that are now constructed are shown in Figures 13.1 and 13.2. These complicated buildings may be difficult to construct because of the difficulties in fabricating complex shapes for the building shell and using new materials in innovative ways. In heavy engineering projects, the scope of the project may make the project difficult to estimate and control. A 20-mile long suspension bridge would be an example of this type of project. Various lifting devices, highly trained labor, and huge bridge sections must be brought together in a cost-efficient way.
- **Use of new materials.** Construction contractors are used to dealing with certain types of materials. The introduction of new material types such as warm-mix asphalt (see Chapter 14) can reduce productivity and increase costs until the contractor learns how to handle the material in the most efficient manner.
- **Location.** Construction costs vary depending on location. A remote site will increase transportation costs of materials, making it difficult to attract skilled labor.

It has also been found that highly urbanized sites such as mid-town Manhattan can be difficult to work in because of limited space and traffic congestion.

- **Quality required.** Specifications define the quality required by the owner. A project having high quality standards will take longer to complete and be costlier than a similar project of lower quality. Companies that normally handle projects with low quality requirements often have difficulty estimating a project requiring high quality.
- **Management factors.** A well-managed project will have better cost performance than a poorly managed project. In a well-managed project, the construction contractor will be aware of project issues and will closely monitor the project to keep costs from exceeding the estimate.

Types of Estimates

This book has shown that a project must travel through several stages before it can be completed. At all stages of a project, various estimates of project cost are used. During the design stage of the project, estimates are made to inform the owner about the expected project cost. If estimates are high, an owner may consider redesigning the project or changing its scope to reduce cost. Later, when the bid has been advertised, construction contractors develop bid estimates. Bid estimates are compared with a final engineer's estimate that is prepared by the designer for the owner. Sometimes, when there are significant differences between the bid prices and the engineer's estimate, an owner may choose to rebid the project.

The level of detail that is possible with an estimate varies depending on how much design information is available. Estimates vary in their type and detail, depending on the stage of the project. Typical types of estimates include (Manfredonia and Majewski, 2008):

- **Order of magnitude.** An **order of magnitude** estimate is a "ballpark" estimate typically made at the initiation of a project before any plans have been drawn. These estimates are based on cost per primary unit of a facility (Jackson, 2004), such as mile per highway or number of condominiums to be constructed.
- **Preliminary estimates.** As the design continues to evolve, preliminary design documents become available for a more detailed estimate. The square foot method of estimating is often used for this type of estimate. Several manuals provide estimates of building cost related to the buildings type in square feet. An estimator can determine the square footage of a project and then multiply by the cost/square foot to determine a preliminary cost number for a project. Figure 8.1 shows a square foot estimate produced by the RS Means CostWorks Web page. Formerly, estimators used printed manuals but with the increasing power of IT, a Web-based program to construct estimates is now available. Figure 8.1 shows how eight inputs are used to determine a **preliminary estimate** of cost for the building of a college dormitory. The area, perimeter, number of stories, and story height are input

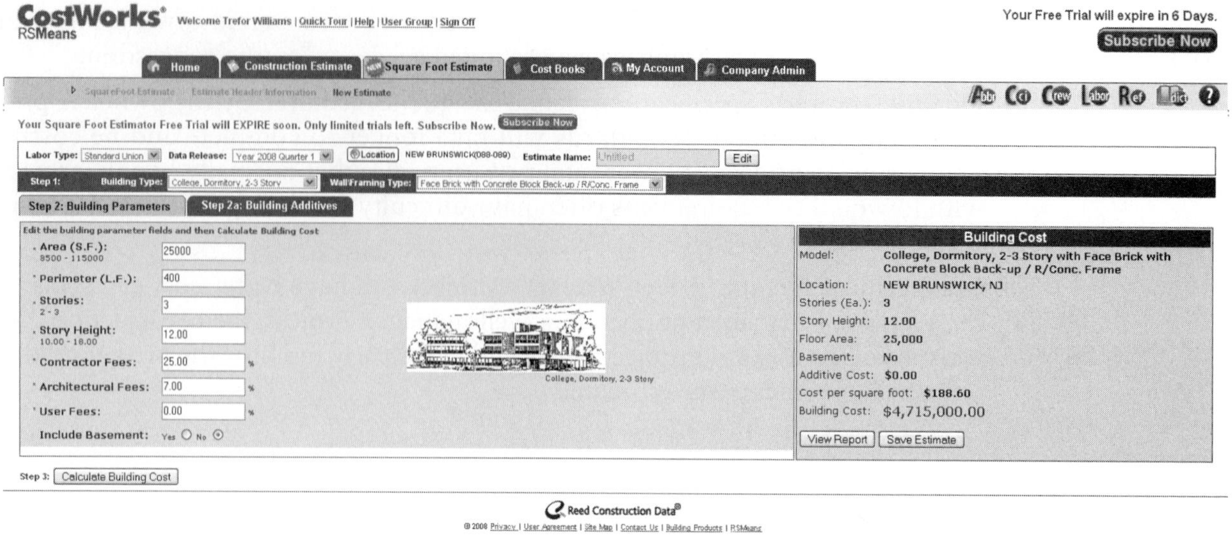

Figure 8.1 Square foot cost estimate for a dormitory. From Means CostWorks Residential Cost Data, 2008. Copyright RSMeans, Kingston, MA 781-585-7880. All rights reserved.

to determine the basic characteristics of the dormitory facility. Contractor and architectural fees are included as a percentage of total cost. The output is shown on the right-hand side of the screen, giving a preliminary cost of $4,715,000.

- **Estimates during design development.** Most owners require several estimates to be calculated by the designer as the design proceeds and more details about the project are known. These estimates confirm that the owner can afford the design, by allowing a value engineering review of the cost of components selected to insure they are within the scope of the projects. Estimates of this type are done with not less than 25% complete drawings and an outline specification.
- **Engineer's estimate.** When plans and specifications are complete, it is possible for the designer to do a detailed **engineer's estimate**. This estimate is a final check for the owner to determine if the project is financially feasible. In addition, the engineer's estimate can be compared with bid estimates submitted by the competing contractors to determine if responsible bids have been received.
- **Bid estimates.** For a competitively bid contract, **bid estimates** must be highly detailed. The bid estimate will include a contractor's profit and overhead. The bid estimate will be adjusted on the basis of the estimator's knowledge of local competition, the competitiveness of the market place, and the company's need to get a project. In periods where few projects are available, contractors may bid with low-profit margins just to keep their firm in business. A desirable bid would be a bid that is lower than all competitors, but not significantly lower. When a low bidder's bid is much lower than the competitor's, it is often called "leaving money on the table."

Estimates and Different Types of Project Delivery Methods

In this book, several types of contract delivery methods have been discussed, in particular, the increasingly popular design-build method. However, a competitively bid design-build project provides a different challenge compared with traditional competitively bid projects. Bidding takes place much earlier in the design-build project, occurring typically when the design is around 20% complete. Therefore, estimates are not as detailed, and the price may be established using preliminary estimating methods. Then, after the low bidder is awarded the project, he or she prepares more detailed estimates as the process of project design progresses.

With negotiated projects, particularly cost plus a fee, a cost estimate is not as important. However, the contractor must have enough information about the project cost to negotiate a guaranteed maximum price. Needless to say, a good preconstruction estimate is important to contractors to establish project budgets.

Components of a Construction Cost Estimate

Construction projects incur many different types of costs. These costs must be recovered to insure project profitability. In addition, the contractor would want to recover his or her overhead cost, which are costs incurred by the contractor but not directly chargeable to any items physically installed on the project.

The **quantity takeoff** is the measurement and calculation from the plans of the quantities of work that need to be performed on a project. For example, earthwork quantities are calculated from the plans for a highway fill, or the quantity of shingles required to roof a house is calculated from the dimensions of the roof. The **unit pricing** process determines how much each unit will cost to produce, transport, and install in the correct position as required by the project (Gould and Joyce, 2002). Material, labor, and equipment costs must be included in the calculation of the unit cost.

The major cost elements of a construction project are (Manfredonia and Majewski, 2008):

- **Labor costs.** These are the costs of project labor, including fringe benefits. Fringe benefits are the additional benefits, such as vacation and health plans, an employee receives. A contractor must also consider the cost of government-mandated programs such as social security and state unemployment insurance. Fringe benefits can be a significant percentage of an employee's salary and must be carefully calculated.
- **Material costs.** Material prices can fluctuate significantly. An estimator requires an understanding of the current economic situation and factors such as exchange rates and material availability to minimize the cost of materials purchased.
- **Equipment costs.** Equipment costs vary based on project conditions and cycle time requirements for equipment balance. Costs differ depending on whether the equipment is owned or leased by the contactor.

- **Subcontractor costs.** Quotes from subcontractors are also a project cost. The subcontractors include their own overhead and profit in the quote.
- **Overhead costs.** There are two types of **overhead costs**: project overhead and direct overhead. Project overhead includes the cost of a field office, supervisory personnel, temporary utility connections, and small tools. Direct overhead, or home office overhead, includes the cost of having a main office, and the support staff in the main office, such as administrative personnel, marketing personnel, and estimators. Sometimes, overhead is applied as a percentage markup of total project cost.
- **Profit.** The contractor's profit must be included in the estimate. Typically, the profit is applied as a percentage of the sum of the total project cost and the overhead cost. The amount of the percentage markup that is applied is guided by the contractor's experience and consideration of factors such as the level of competition, the need for new work, economic conditions, and the availability of new projects.

Preliminary Estimating Using Cost Indexes

For preliminary cost estimates, a cost index may be used to project construction prices from the past to the present or future. Construction companies maintain databases of the costs of past projects. Past project costs will not be directly equivalent to present prices because of the effects of inflation. A **cost index** is used to bring the price to a present or future value. There are standard published cost indexes, and it is possible to build a customized index for a construction company.

Example

ENR Cost Indexes

Several well-known construction cost indexes are maintained. Perhaps the most well known are the ENR Building Cost and Construction Cost Indexes. The Building Cost Index is a 20-city average market basket of construction. It consists of 63.38 hours of skilled labor at the 20-city average of bricklayer, carpenter, and structural ironworker rates, plus 25 cwt of standard structural steel shapes at the mill price prior to 1996 and the fabricated 20-city price from 1996, plus 1.128 tons of portland cement at the 20-city price, plus 1,088 board-ft of 2 × 4 lumber at the 20-city price. The Construction Cost Index consists of 200 hours of common labor at the 20-city average of common labor rates, plus 25 cwt of standard structural steel shapes at the mill price prior to 1996 and the fabricated 20-city price from 1996, plus 1.128 tons of portland cement at the 20-city price, plus 1,088 board-ft of 2 × 4 lumber at the 20-city price (ENR, 2009). The value of the indexes varies because of changes in cost of its constituents.

This example will illustrate how an index can be used to project costs. Table 8.1 shows Construction Cost Index values prepared by a construction company on the basis of its own project costs. Economists working for the firm have projected the index values into the future. The initial year of the index was 2005, which was given an arbitrary value of 100.

Table 8.1 A typical Construction Cost Index

YEAR	CONSTRUCTION COST INDEX VALUES
2005	100
2006	109
2007	86
2008	123
2009	130
2010	137
2011	142
2012	151
2013	159
2014	164

The company built a project in 2006 that is similar to the proposed project due to be built in 2014. The project in 2006 was completed for a cost of $1,250,000. The following formula gives the projected cost for the new project in 2014:

$$C_f = C_p\left(\frac{I_f}{I_p}\right)$$

where C_f is the future or present inflated price of a project, C_p is the past cost of a similar project, I_f is the future value of the index, and I_p is the past value of the Construction Cost Index. In this example, the 2014 price would be calculated as:

$$\$1{,}250{,}000 \times 164/109 = \$1{,}880{,}734.$$

Detailed Estimating Using an Online Reference Manual

Figure 8.2 shows a list of items published in the RSMeans Web-based software called CostWorks. Notice for each detailed line item, CostWorks provides the unit of measure, the crew composition, and the daily output. Bare costs (costs that do not include overhead and profit) are provided for labor, materials, and equipment. The total base cost shown is the sum of the bare costs of labor, material, and equipment. The last column provides the total cost plus overhead and profit. The costs included in the RSMeans reference are obtained nationally. Users can choose to use a national average price or select the specific geographic area they are working in.

An example of how to use the RSMeans system is given in Figure 8.2. Line number 057113501810 gives the cost for the installation of a 5' diameter spiral aluminum

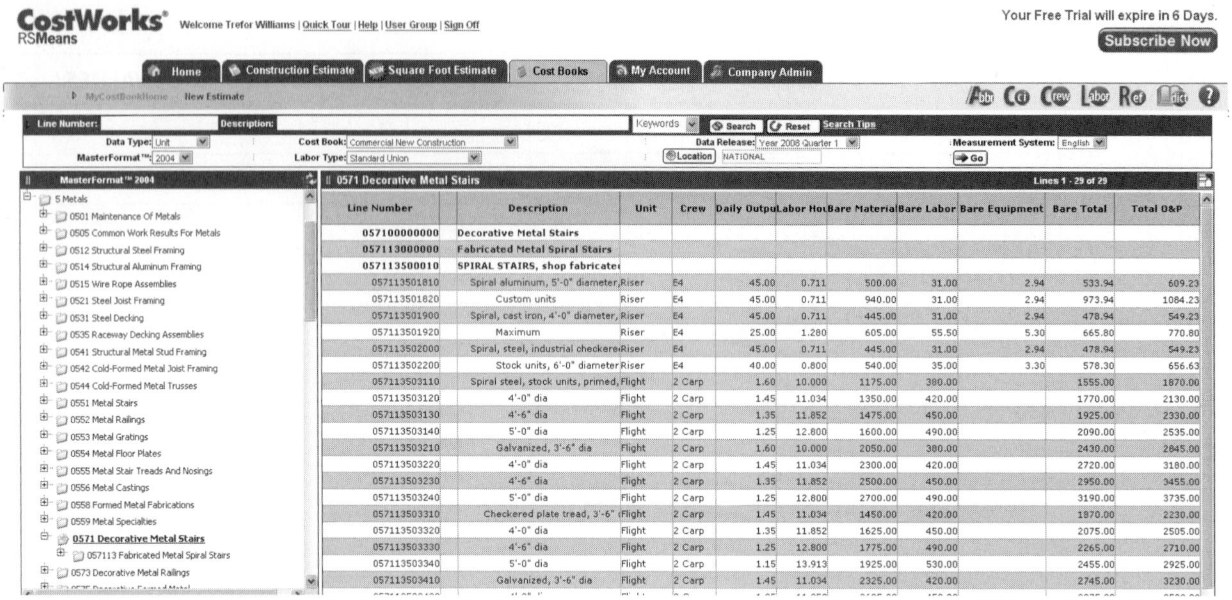

Line Number	Description	Unit	Crew	Daily Outpu	Labor Hou	Bare Materia	Bare Labor	Bare Equipment	Bare Total	Total O&P
057100000000	Decorative Metal Stairs									
057113000000	Fabricated Metal Spiral Stairs									
057113500010	SPIRAL STAIRS, shop fabricate									
057113501810	Spiral aluminum, 5'-0" diameter,	Riser	E4	45.00	0.711	500.00	31.00	2.94	533.94	609.23
057113501820	Custom units	Riser	E4	45.00	0.711	940.00	31.00	2.94	973.94	1084.23
057113501900	Spiral, cast iron, 4'-0" diameter,	Riser	E4	45.00	0.711	445.00	31.00	2.94	478.94	549.23
057113501920	Maximum	Riser	E4	25.00	1.280	605.00	55.50	5.30	665.80	770.80
057113502000	Spiral, steel, industrial checkere	Riser	E4	45.00	0.711	445.00	31.00	2.94	478.94	549.23
057113502200	Stock units, 6'-0" diameter	Riser	E4	40.00	0.800	540.00	35.00	3.30	578.30	656.63
057113503110	Spiral steel, stock units, primed,	Flight	2 Carp	1.60	10.000	1175.00	380.00		1555.00	1870.00
057113503120	4'-0" dia	Flight	2 Carp	1.45	11.034	1350.00	420.00		1770.00	2130.00
057113503130	4'-6" dia	Flight	2 Carp	1.35	11.852	1475.00	450.00		1925.00	2330.00
057113503140	5'-0" dia	Flight	2 Carp	1.25	12.800	1600.00	490.00		2090.00	2535.00
057113503210	Galvanized, 3'-6" dia	Flight	2 Carp	1.60	10.000	2050.00	380.00		2430.00	2845.00
057113503220	4'-0" dia	Flight	2 Carp	1.45	11.034	2300.00	420.00		2720.00	3180.00
057113503230	4'-6" dia	Flight	2 Carp	1.35	11.852	2500.00	450.00		2950.00	3455.00
057113503240	5'-0" dia	Flight	2 Carp	1.25	12.800	2700.00	490.00		3190.00	3735.00
057113503310	Checkered plate tread, 3'-6" (	Flight	2 Carp	1.45	11.034	1450.00	420.00		1870.00	2230.00
057113503320	4'-0" dia	Flight	2 Carp	1.35	11.852	1625.00	450.00		2075.00	2505.00
057113503330	4'-6" dia	Flight	2 Carp	1.25	12.800	1775.00	490.00		2265.00	2710.00
057113503340	5'-0" dia	Flight	2 Carp	1.15	13.913	1925.00	530.00		2455.00	2925.00
057113503410	Galvanized, 3'-6" dia	Flight	2 Carp	1.45	11.034	2325.00	420.00		2745.00	3230.00

Figure 8.2 CostWorks data for spiral stairs. From Means CostWorks Residential Cost Data, 2008. Copyright RSMeans, Kingston, MA 781-585-7880. All rights reserved.

staircase. Note that this line number conforms to the CSI system and the first four digits designate decorative metal stairs. Additional information can be quickly calculated from the line item. The time to complete a line item is given by:

$$\text{Quantity of work/daily output from line item} = \text{Time to complete}$$

If 180 risers must be constructed, the time to perform the work is calculated as:

$$180 \text{ risers}/45 \text{ risers/day} = 4 \text{ days}$$

The CostWorks line item provides the labor hours to complete one riser as 0.78 labor hours/riser. Total labor hours for a project bid item are given as:

$$\text{Labor hours/unit} \times \text{Number of units} = \text{Total labor hours}$$

Therefore the labor hours for 180 risers is:

$$0.78 \text{ labor hours/riser} \times 180 \text{ risers} = 127.98 \text{ labor hours}$$

The total cost of the spiral staircases including overhead and profit is found by multiplying the last column of the item by the total quantity of risers:

$$\$609.23/\text{riser} = 180 \text{ risers} = \$109{,}661.40$$

We can see from this example that the RSMeans CostWorks provides rapid access to productivity and cost data for thousands of different line items. It is particularly useful to estimators who have no historical records of the type of work being estimated.

Figure 8.3 Setting up a detailed Bid in CostWorks. From Means CostWorks Residential Cost Data, 2008. Copyright RSMeans, Kingston, MA 781-585-7880. All rights reserved.

Figure 8.3 shows the form used to initiate a detailed estimate using CostWorks. Figure 8.4 shows how line items are selected for inclusion in the estimate. The user needs only to select an item in the top pane and click the add button to add it to the estimate. The user inputs the required quantity and CostWorks automatically calculates the total line item cost and the total project cost.

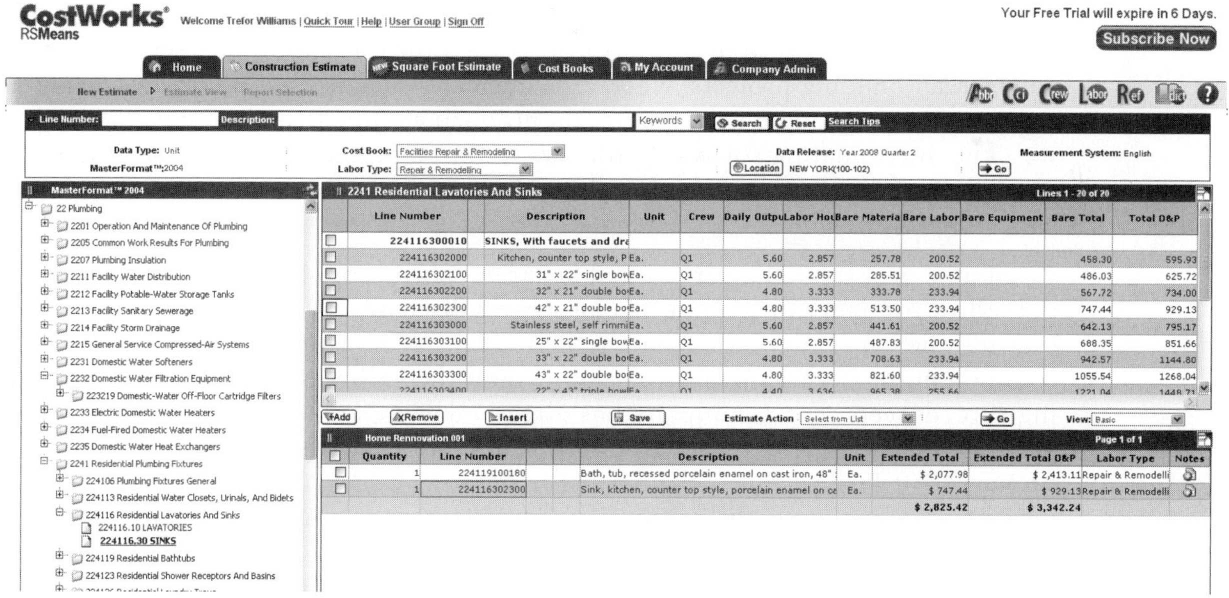

Figure 8.4 Preparing a detailed estimate. From Means CostWorks Residential Cost Data, 2008. Copyright RSMeans, Kingston, MA 781-585-7880. All rights reserved.

Detailed Estimating Using Computer Software

Estimating using computer software is rapidly replacing pen-and-paper-based estimates. The primary reasons are the reduction of errors that are possible using the computer and reduced estimating costs. Many computer programs are available to assist in developing estimates for construction projects.

More opportunities to automate the estimating process are emerging. It is now possible to generate data automatically through the use of interoperability standards with CAD documents. CAD documents can be coded with detailed information about each project item that can be transferred to estimating programs (see Chapter 13). Additionally, programs are available that allow for automated quantity takeoff from electronic plan sheets. This provides the capability for a "paperless" estimating process with reductions in printing costs and time spent entering data manually.

Many estimating software programs are available. The cost of the estimating software varies from under $100 to many thousands of dollars. There are many programs that have been developed for small construction projects and homebuilders. Well known among the software developed for small projects are Goldenseal, Bid4Build, and WinEstLT. Typically, the software for small projects does not have the sophisticated data exchange capabilities or integration capabilities with other software packages that is contained in software conceived to handle large projects. Additionally, these software packages may be limited in their ability to develop the complex item coding schemes that may be used on a large project. However, they are ideal for a small contractor because they are easy to learn and use.

For sophisticated projects, the different software packages tend to concentrate in particular segments of the construction market. For example, Timberline Office is a powerful estimating software that is most often used for estimating the construction of commercial building projects, whereas a software such as HeavyBid is more suited to infrastructure projects and focuses on producing a line item, unit cost bid. Other well-known estimating software products for large projects are Hard Dollar, MC^2, and Bid2Win. Bid2Win and Hard Dollar are more oriented toward highway and infrastructure projects whereas MC^2 is aimed at commercial construction.

Computer programs that are aimed at large contractors have powerful features that allow for data integration. The programs discussed in preceding text typically provide the capability to exchange data with Excel. The use of spreadsheets is common in the development of estimates, and the ability to use standard estimating spreadsheets that a company has already developed is a useful feature. Most programs have the capability to export data to Primavera Project Planner and/or Microsoft Project. Both Timberline Estimating and Primavera are compatible with the IFC **interoperability** specifications (International Alliance for Interoperability). This allows for a sophisticated linking between the two programs. Timberline Estimating includes a module called scheduling integrator that allows data from the Timberline estimate to be transferred to the Primavera scheduling program to automatically create scheduling activities. Activity durations are automatically generated from the estimate data. The generated activities can be automatically grouped in different ways, including by work breakdown structure, phase, or location (Timberline Software Corporation, 2004c).

Some estimating software, including Timberline Estimating and MC^2, provide built-in modules that allow paperless methods of quantity takeoff by automatically calculating quantities from scaled CAD documents. Timberline Estimator has a module called On-Screen Takeoff. The software allows scaled drawings in different CAD and graphic formats to be used to receive takeoff values and transfer them into Timberline estimating spreadsheets. The On-Screen Takeoff module allows lengths, areas, and volumes to be transferred for use in Timberline estimating. File formats that can be used include DWG, DXF, JPEG, BMP, TIFF, and PDF. The On-Screen Takeoff software also accepts other types of files, including Dodge Plan format, and government formats such as the Corps of Engineers. On-Screen Takeoff can also produce Excel files as output (Timberline Software Corporation, 2004a).

On-Screen Takeoff (On Center Software, 2005) can be purchased separately for use with other estimating software packages. Figure 8.5 shows a landscaping CAD file that has been loaded into the On-Screen Takeoff Program image window. Various objects on the plan have been designated as different project items. In this case,

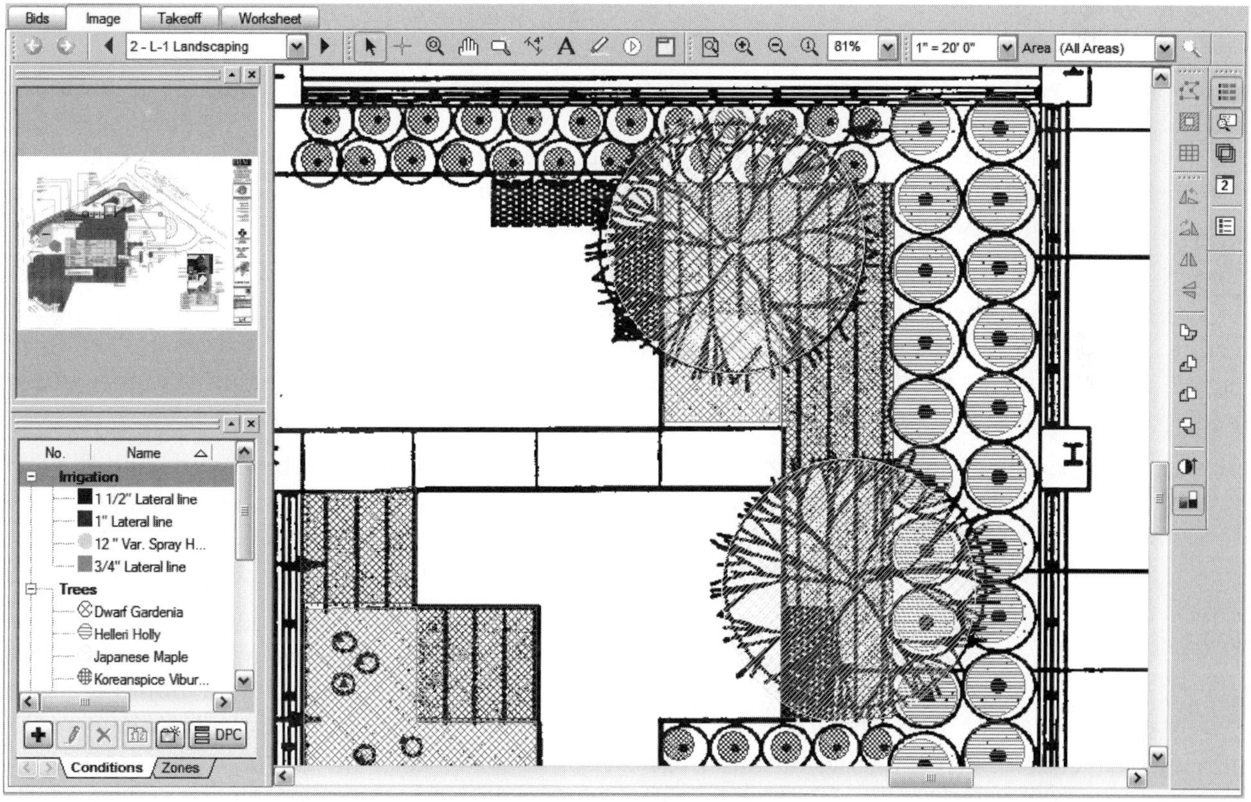

Figure 8.5 Computerized quantity takeoff. *Courtesy of On Center Software, Inc.*

Bids | Image | Takeoff | Worksheet

No.	Name	Height	Area	Quantity 1	UOM1	Quantity 2	UOM2	Quantity 3	UOM3	Notes
Bed Plants										
5	Variegated Liriope	0"	(unassigned)	1,293	SF	321	LF	4	EA	Variegated Lir...
6	Big Blue Monkey Grass	0"	(unassigned)	1,403	SF	383	LF	5	EA	Big Blue Mon...
Ground Cover										
2	Mulch	0"	(unassigned)	3,761	SF	504	LF	2	EA	
3	Grass	0"	(unassigned)	276	SY	2,480	SF	196	LF	
Irrigation										
13	3/4" Lateral line	0"	(unassigned)	57	LF	8	EA	0	SF	3/4" Lateral li...
14	1" Lateral line	0"	(unassigned)	48	LF	8	EA	0	SF	1" Lateral line
15	1 1/2" Lateral line	0"	(unassigned)	21	LF	3	EA	0	SF	1 1/2" Lateral...
16	12 " Var. Spray Head	0"	(unassigned)	21	EA	21	EA	21	EA	12" variable s...
Miscellaneous										
1	Edge Trim	0"	(unassigned)	222	LF	106	EA	0	SF	
Trees										
7	Nellie R Stevens Holly	0"	(unassigned)	9	EA	9	EA	9	EA	
8	Japanese Maple	0"	(unassigned)	5	EA	5	EA	5	EA	
9	Peacock Orchid	0"	(unassigned)	21	EA	21	EA	21	EA	Peacock Orc...
10	Helleri Holly	0"	(unassigned)	36	EA	36	EA	36	EA	Helleri Holly 3...
11	Dwarf Gardenia	0"	(unassigned)	44	EA	44	EA	44	EA	Dwarf Garden...
12	Koreanspice Viburnum	0"	(unassigned)	1	EA	1	EA	1	EA	Koreanspice ...

Figure 8.6 Automatically calculated takeoff quantities. *Courtesy of On Center Software, Inc.*

various plants, trees, and irrigation equipment have been designated. Figure 8.6 shows the On-Screen Takeoff window for the plan sheet. Here, items that have been selected in the image window are automatically tabulated. On-Screen Takeoff is capable of automatically counting, determining linear dimensions, and calculating volumes and areas.

The Capture Desktop feature of MC^2 provides similar functions and allows drawing to be sent in the .jpg format attached to e-mails when there are questions about plan details. This can quicken the exchange of information between the field and office or the designer when there are construction problems in the field (Management Computer Controls, Inc., 2007).

Some of the estimating software include standard cost database information. The Timberline estimating software includes databases of cost information such as the RS Means building construction data. The inclusion of these databases allows a user to produce estimates rapidly even if they do not have their own historical data to draw on.

Software that focuses on commercial building construction tends to have some parametric estimating capabilities. MC^2 includes "estimating wizards" that allow conceptual estimates to be created. The wizards use a typical office building as an example and build a detailed estimate on the basis of the input of the building parameters such as dimensions and number of rooms. Timberline Estimating contains the Timberline Office Commercial Knowledgebase that contains pre-built models and assemblies of various construction systems (Timberline Software Corporation, 2004b).

Table 8.2 Links to construction estimating software

ESTIMATING SOFTWARE	WEB PAGE	COMMENTS
Bid4Build	www.bid4build.com	Small projects
Goldenseal	www.turtlesoft.com	Small projects
WinEst	www.winest.com	Commercial construction
MC^2	www.mc2-ice.com	Commercial construction
Timberline	http://www.sagecre.com/products/timberline_office/estimating	Commercial construction
Hard Dollar	www.harddollar.com	Heavy construction
Bid2Win	www.bid2win.com	Heavy construction
HeavyBid	www.hcss.com	Heavy construction
On-Screen Takeoff	www.oncenter.com/products/ost.asp	Quantity takeoff software

The Commercial Knowledgebase allows a user to answer questions on a form, and then a preliminary or detailed estimate is produced automatically. The purpose of the Commercial Knowledgebase is to allow users to prepare conceptual estimates quickly by eliminating the process of individual item takeoff. The user is prompted only for the data needed to calculate quantities and costs for assemblies or entire buildings. URLs for all the software packages discussed in this section are given in Table 8.2.

An Example Project Using HeavyBid

HeavyBid is an estimating computer program that focuses on the heavy construction segment of the construction market. It focuses on the infrastructure market and provides five levels of the HeavyBid program depending on the size of the project. HeavyBid has been used on projects from $50,000 to over $1 billion. To illustrate the functions of estimating software and to show how a bid is prepared, several examples from the HeavyBid program will be presented. Examples that outline how a bid is prepared using HeavyBid are presented. Following is an example HeavyBid application that also illustrates many of the issues in producing a detailed bid estimate.

The Master Estimate

A master estimate can be established for use with HeavyBid. In this master estimate, the various types of labor, equipment, materials, and subcontractors can be defined. The information entered in the master estimate is information that is used repeatedly in the construction firm's project. Thus, the master estimate can be used as the basis for all new project estimates. HeavyBid also allows any other existing estimate to be used as the basis for a new estimate. If the construction company has just purchased a

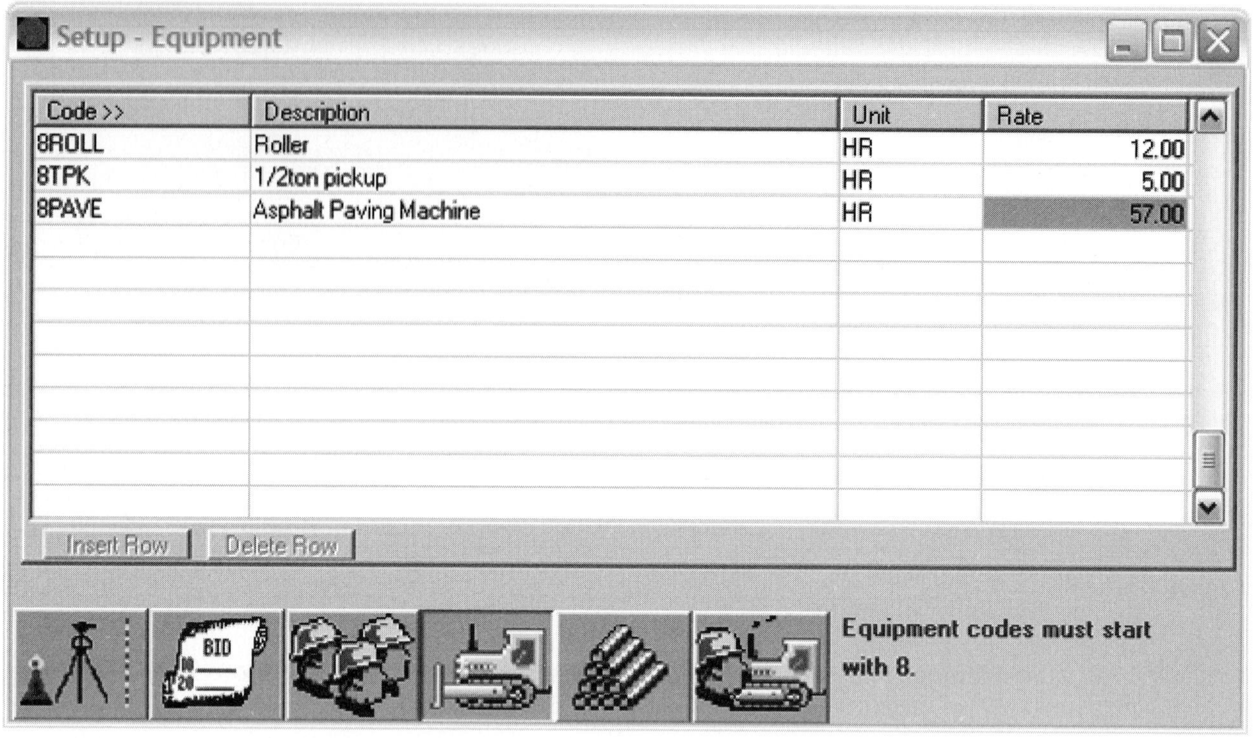

Figure 8.7 Adding equipment to master estimate. *Courtesy of HCSS.*

new asphalt-paving machine for use on its projects, we can add it to the master schedule by clicking on the equipment icon. This is the icon with a picture of a bulldozer. Clicking on this icon opens a window where we can input the new asphalt-paving machine and its charging rate. Figure 8.7 shows how the new machine is added to the master estimate in HeavyBid.

Establishing a Bid

HeavyBid takes a hierarchical approach to establishing project costs. It develops a bid price for each bid item that is defined for a project. Infrastructure projects are typically unit price and require a unit price and total price to be developed for each bid item. Each bid item is composed of activities that consume resources. Each activity requires various resources such as labor, equipment, materials, and subcontractors to complete. These various resources and their costs are associated with each activity and the activity costs are aggregated to develop the bid item price.

HeavyBid has the capability of reading in standard bid items from DOTs or the bid items can be entered manually. Figure 8.8 shows the list of bid items for an asphalt-paving project that are listed in a tree structure at the left of the screen. If a bid item is clicked, the tree expands to show the list of activities incorporated in the item. If an activity is selected, the tree expands further to show all the resources required for the activity. In this case, we see a listing of the equipment and labor required for asphalt

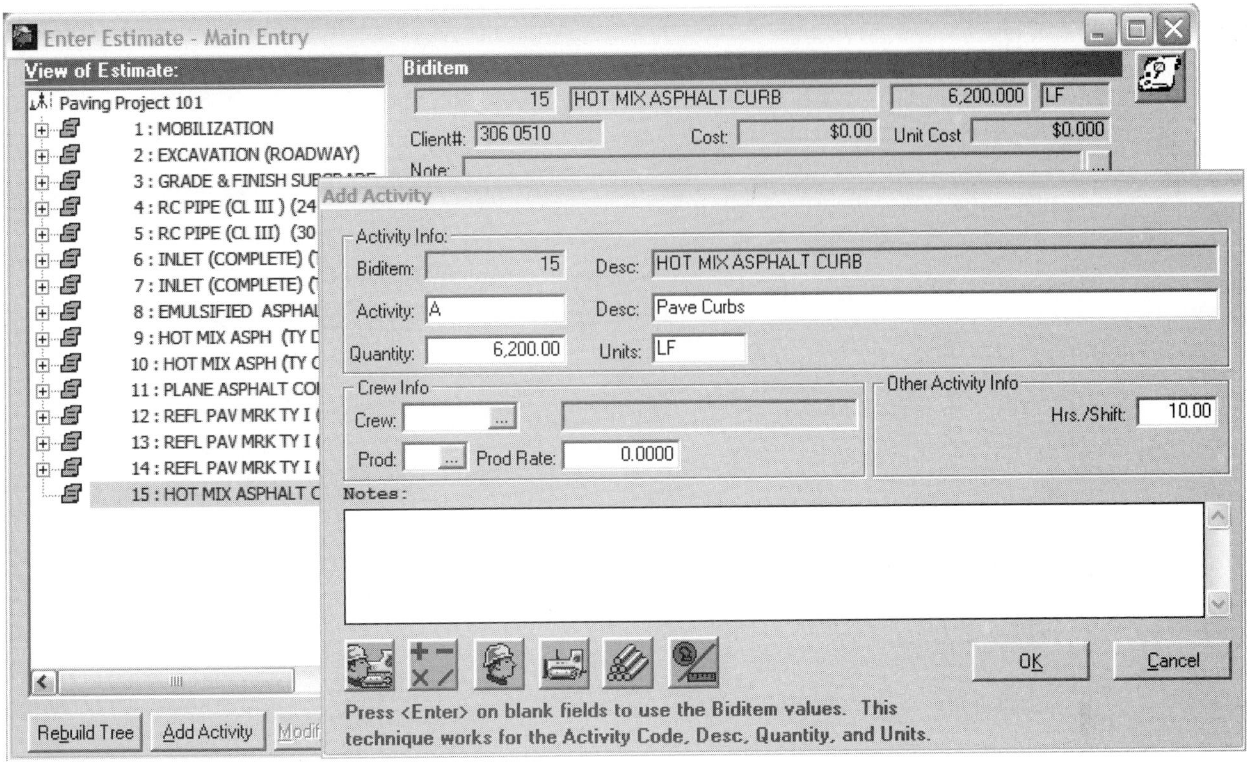

Figure 8.8 Bid item tree. *Courtesy of HCSS.*

paving. Figure 8.9 shows the asphalt-paving bid item completely expanded, showing all the activities, and then a listing of the resources required for an activity.

In HeavyBid, the project is broken up into a hierarchy of items and activities that consist of the required materials, labor, and equipment for the project. This hierarchical approach assists in the identification of the required resources. A bid item is decomposed into its constituent activities. This allows the estimator to then identify the required resources for each activity.

Additional bid items can be easily added to an estimate. Bid items in HeavyBid are numbered. The user is required to input the quantity and unit for the bid item. After adding a bid item, the user goes to the main entry screen for the bid and selects the new bid item. To add an activity to the bid item, we click on the Add Activity tab in the lower left of the screen and the Add Activity window appears. This is illustrated in Figure 8.10. The HeavyBid instructions indicate that activities should be consecutively lettered for small projects. Therefore, the activity is given the designation "A." Any name can be input for an activity; we have used "Pave Curbs" here. We could use the predefined crews to define this activity but instead we will select the required resources from the lists of defined resources.

Equipment, labor, and material resources can be easily added to an estimate in HeavyBid. These windows are activated by clicking on the icons at the right of the estimate window.

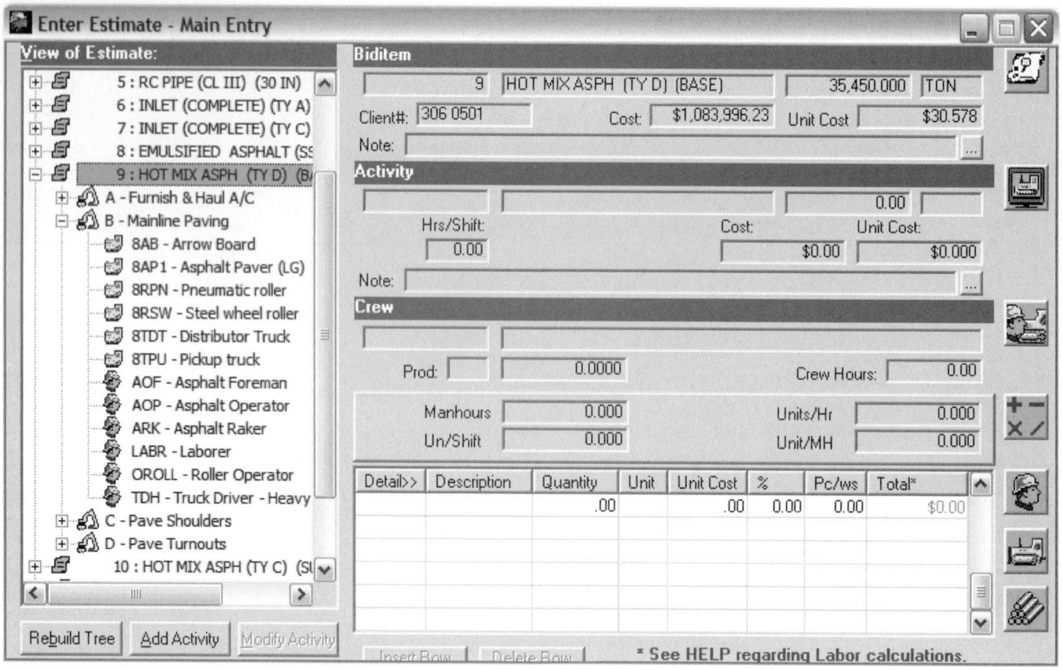

Figure 8.9 Expanded tree showing activities and resources. *Courtesy of HCSS.*

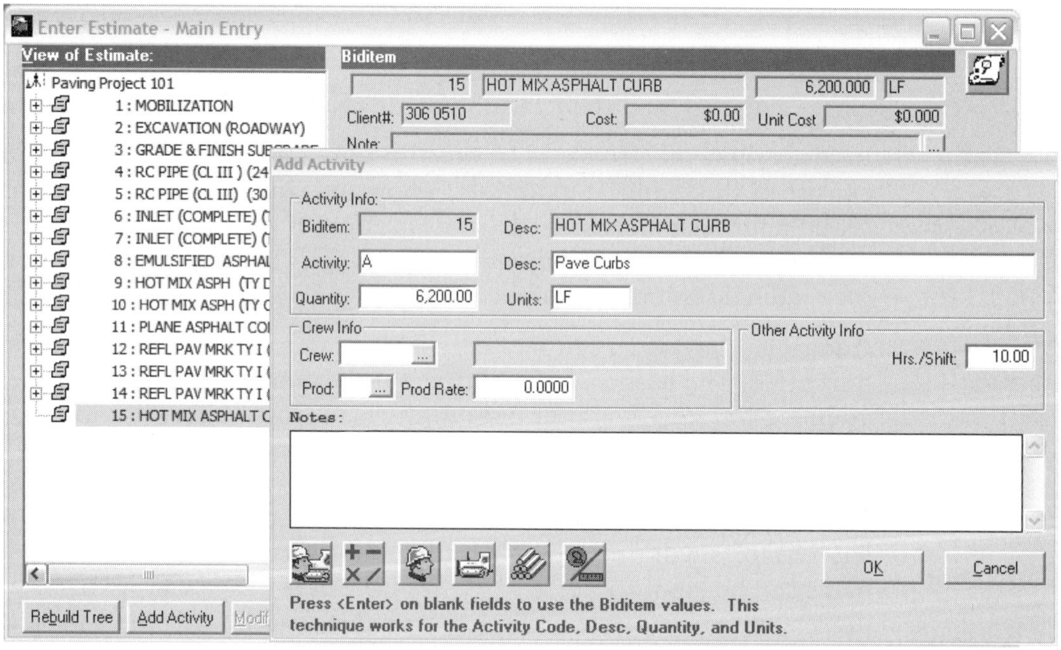

Figure 8.10 Adding an activity. *Courtesy of HCSS.*

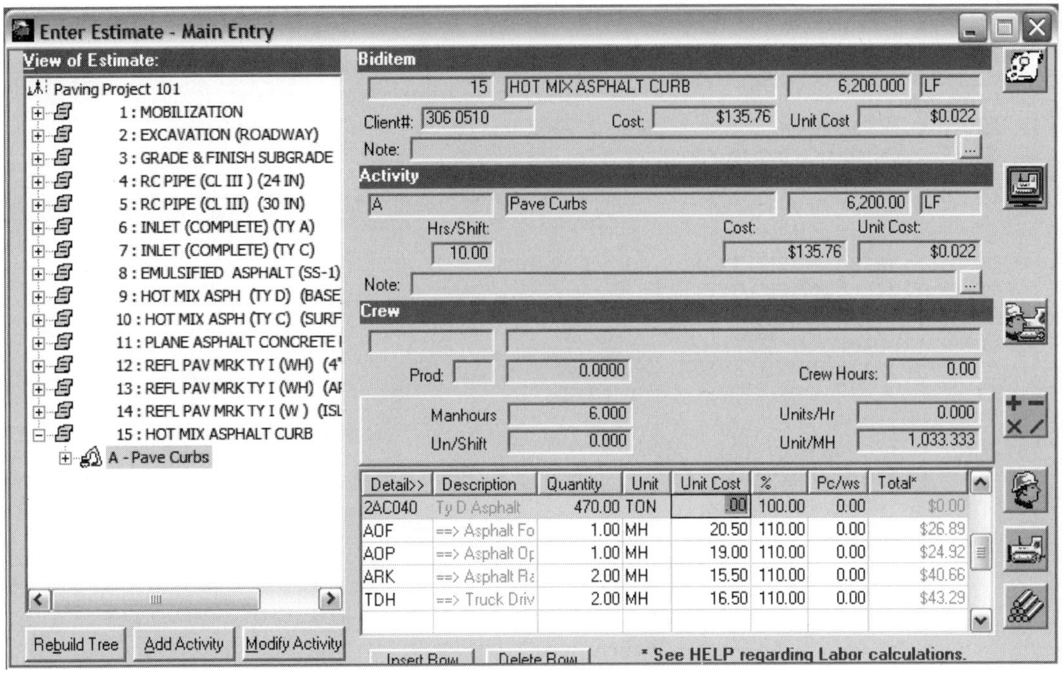

Figure 8.11 Entering material unit price. *Courtesy of HCSS.*

The construction worker icon is for labor, and the bulldozer icon is for equipment. Selecting the icon resembling pipes brings up the predefined materials list. In this case, we select hot mix asphalt and it is brought into the activity. Figure 8.11 shows the final step of entering the material information where the unit price for the asphalt is entered in the activity window.

Finally, to complete the addition of the activity, we adjust the man hours required for the labor and equipment on the basis of our knowledge of the time required to complete the activity. This completes adding the bid item to the estimate (Figure 8.12). Note that all the details of the added resources are shown in the spreadsheet at the bottom right of the screen. It should also be noted that Crew section of the activity is blank. If we had used standard crews, this information would be filled in. It is important to note that the productivity of a crew is defined in this window. Various units for productivity can be selected from the productivity code window. These include units/hr, units/shift, units/man-hour, man-hours/unit, $/unit, and crew hours.

The HeavyBid program also allows new resource information to be inserted directly into the activity spreadsheet without the need to predefine the resources. For example, an equipment rental not part of a company's equipment fleet could be entered directly for a specific project.

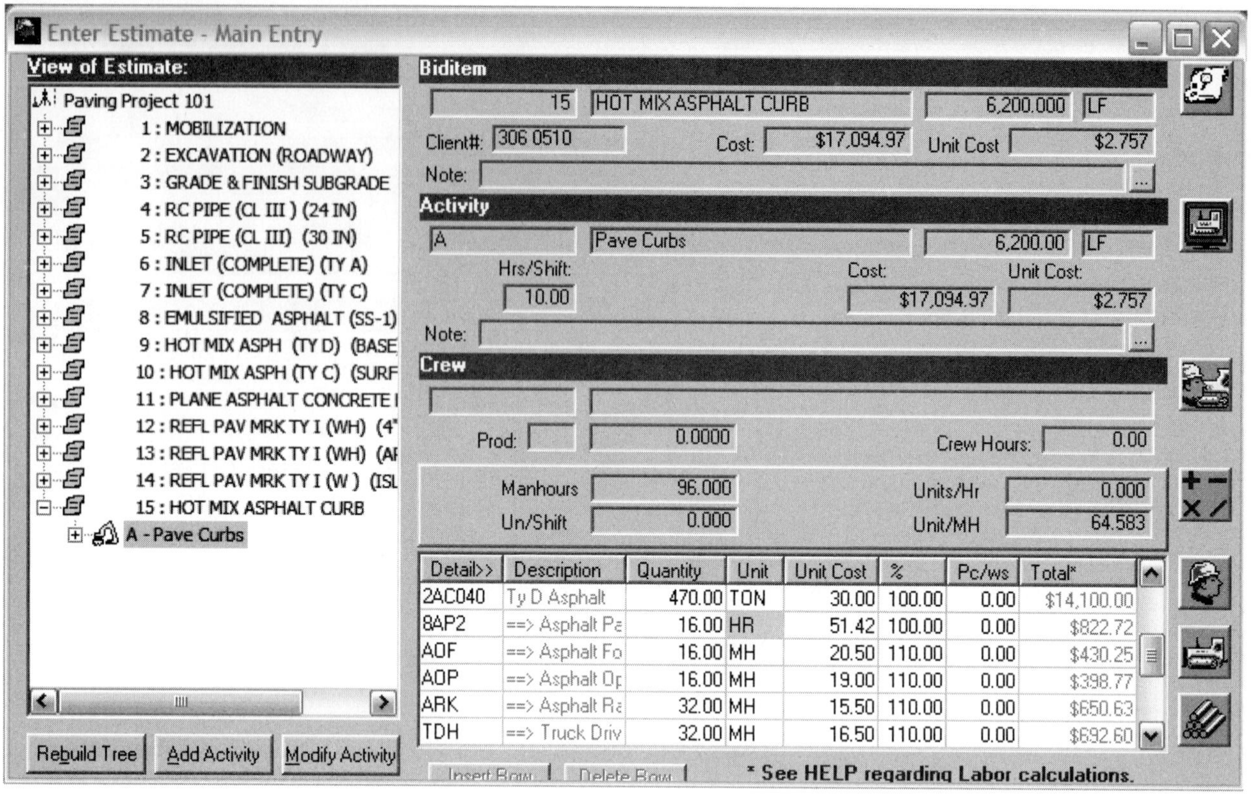

Figure 8.12 Completed activity with man-hours adjusted. *Courtesy of HCSS.*

Finishing the Bid Estimate

Entering all the bid items and their associated activities produces the contractor's cost to perform the work. Of course, the bid estimate submitted to the owner must include markups for overhead and profit. A strength of HeavyBid is its ability to produce the marked up costs that are necessary for the unit price type of contracts seen in infrastructure projects.

Figure 8.13 shows the bid summary screen. Here, a contractor may determine a separate percentage markup for each type of cost. Additionally, a contractor may have certain costs that he or she wishes to spread over all project items. By clicking on the Enter Addons tab of the Bid Summary window, it is possible to add these additional costs. Figure 8.14 shows the add-ons for this project. In this case, overheads are going to be added to the project cost and spread over all the bid items' total direct costs. Additional options are also possible, including spreading the cost over labor costs only, prorating the costs over total costs less subcontractor costs, or no spread. If no spread is selected, the additional costs must be distributed by using manual bid unbalancing. The markup is spread back to the bid items in a similar manner. To calculate the total contract cost after markups and additional costs are included, the user must click on

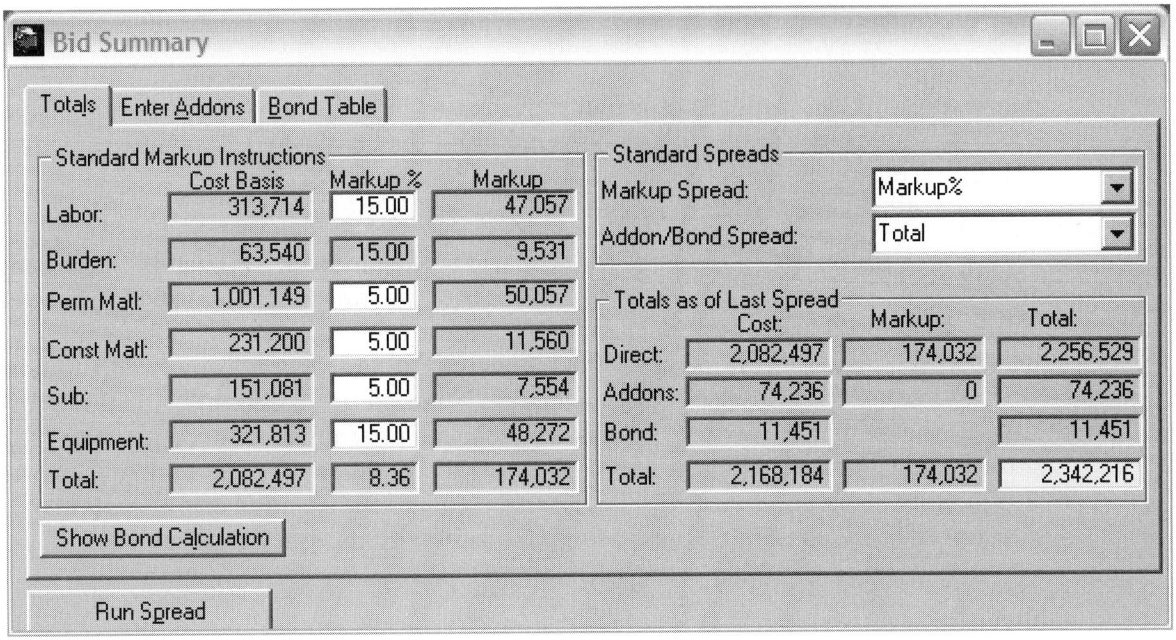

Figure 8.13 Bid summary screen. *Courtesy of HCSS.*

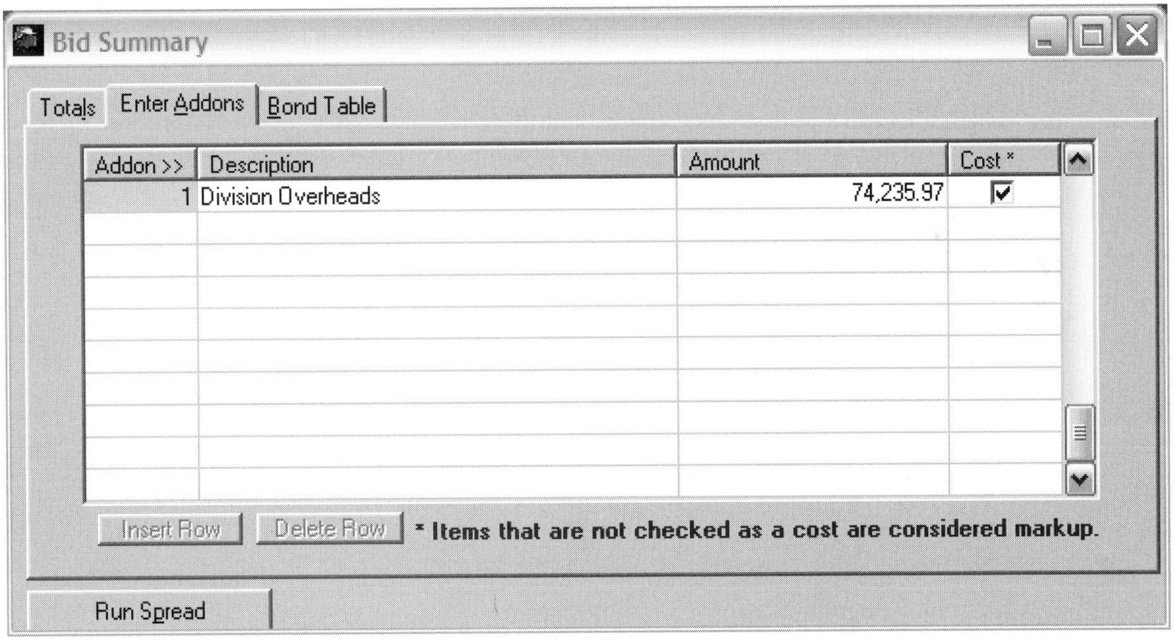

Figure 8.14 Project add-ons. *Courtesy of HCSS.*

the Run Spread button. The total project cost is seen in the bottom right-hand corner of the bid summary screen.

The cost of the bid bonds is also incorporated in the bid at this point. By clicking on the Bond Table tab, it is possible to view and input the bonding costs associated with the project. Typically, bond prices are quoted as a percentage of the contract amount, and this table allows these percentages to be easily input.

Figure 8.15 shows the Bid Pricing window where the bid price for each bid item is displayed. This is the final stage of the bid preparation process using HeavyBid. This screen shows a balanced price, bid price, and bid total for each item. Each bid item has a balanced price; this is the unit price for the bid item including all direct costs and additional overhead costs. The bid price is the same as the balanced price, but can be modified manually by the estimator. The bid total column is the bid price times the item quantity. Therefore, it can be observed that the HeavyBid program automatically takes the raw cost and productivity input, and transforms it automatically into the required unit prices and totals that would be required to fill out a proposal form for an infrastructure project bid.

HeavyBid also has the capability of checking for bid errors. Figure 8.16 shows the Estimate Inquiry screen that is accessed from the query menu. This checking of the bid

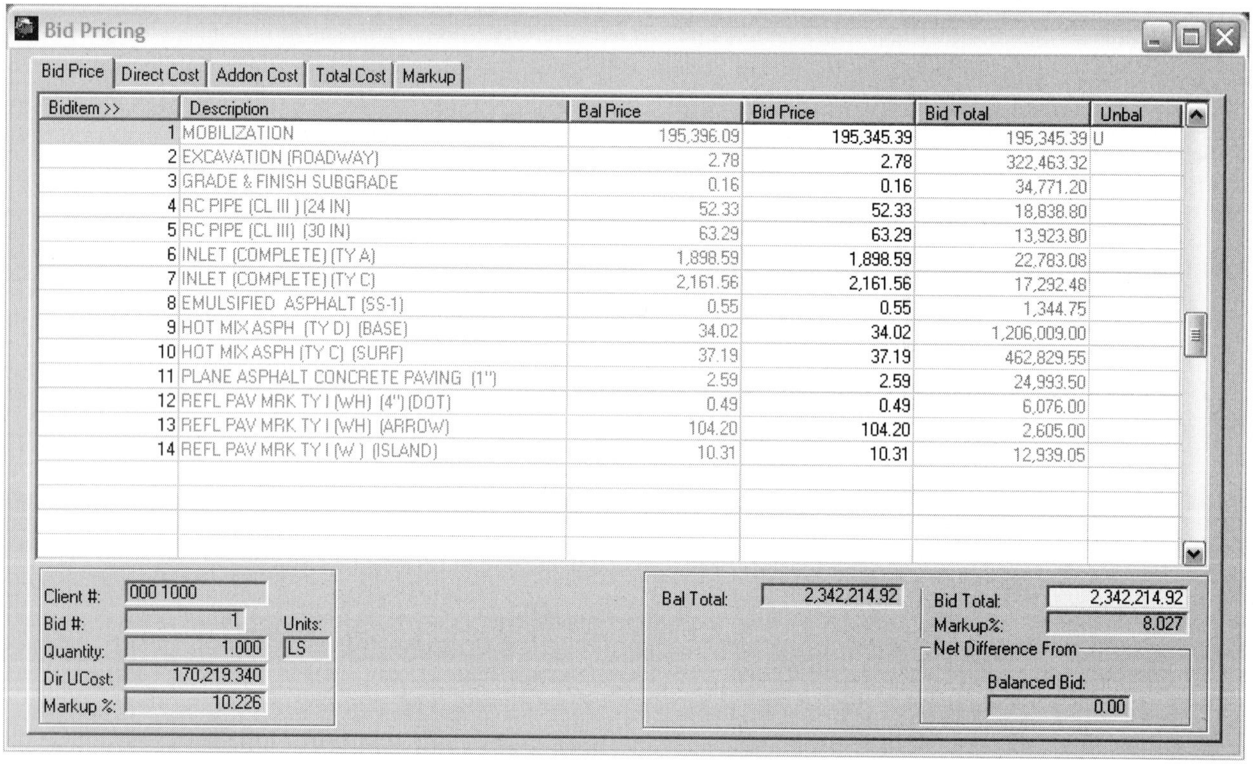

Biditem >>	Description	Bal Price	Bid Price	Bid Total	Unbal
1	MOBILIZATION	195,396.09	195,345.39	195,345.39	U
2	EXCAVATION (ROADWAY)	2.78	2.78	322,463.32	
3	GRADE & FINISH SUBGRADE	0.16	0.16	34,771.20	
4	RC PIPE (CL III) (24 IN)	52.33	52.33	18,838.80	
5	RC PIPE (CL III) (30 IN)	63.29	63.29	13,923.80	
6	INLET (COMPLETE) (TY A)	1,898.59	1,898.59	22,783.08	
7	INLET (COMPLETE) (TY C)	2,161.56	2,161.56	17,292.48	
8	EMULSIFIED ASPHALT (SS-1)	0.55	0.55	1,344.75	
9	HOT MIX ASPH (TY D) (BASE)	34.02	34.02	1,206,009.00	
10	HOT MIX ASPH (TY C) (SURF)	37.19	37.19	462,829.55	
11	PLANE ASPHALT CONCRETE PAVING (1")	2.59	2.59	24,993.50	
12	REFL PAV MRK TY I (WH) (4") (DOT)	0.49	0.49	6,076.00	
13	REFL PAV MRK TY I (WH) (ARROW)	104.20	104.20	2,605.00	
14	REFL PAV MRK TY I (W) (ISLAND)	10.31	10.31	12,939.05	

Client #: 000 1000
Bid #: 1
Quantity: 1.000 Units: LS
Dir UCost: 170,219.340
Markup %: 10.226

Bal Total: 2,342,214.92

Bid Total: 2,342,214.92
Markup%: 8.027
Net Difference From Balanced Bid: 0.00

Figure 8.15 Bid pricing window. *Courtesy of HCSS.*

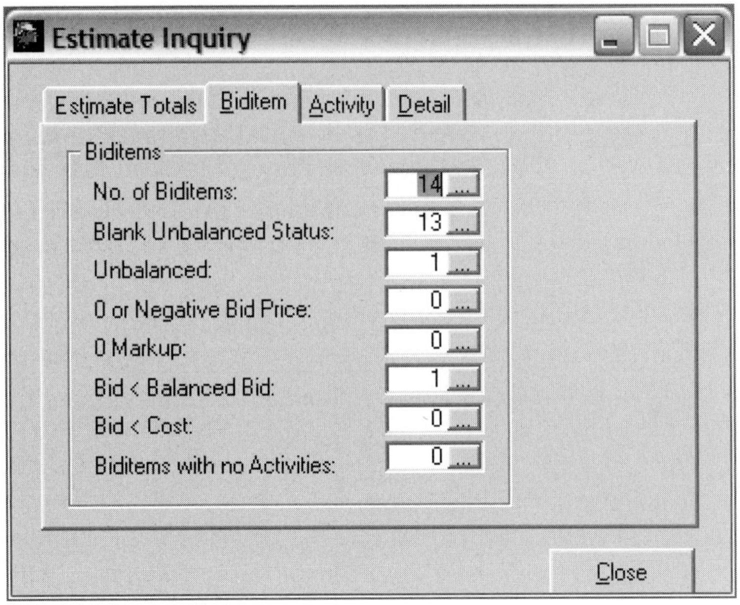

Figure 8.16

Estimate inquiry window.
Courtesy of HCSS.

would be difficult using a paper-and-pencil bidding technique. Using computer software can significantly reduce bidding errors. In this screen, possible bidding errors are highlighted. For example, this screen will display the number of items where the bid price is less than the calculated cost. The screen also flags bid items that are zero, probably indicating a bid item that has not been populated with cost data and is blank.

Summary

In this chapter, it has been demonstrated how knowledge of work task productivity, quantity of materials, labor, and equipment to be used can be combined to determine the work task's cost. Estimates are required throughout the project life cycle and vary in complexity from simple parametric estimates to detailed line item estimates. Computers are now being widely used for construction estimating and this chapter demonstrated their use and application. To achieve profitability, an estimate must include the many costs incurred on a construction project plus markups for overhead and profit.

Key Terms

Bid estimates
Cost index
Engineer's estimate
Estimate
Interoperability
Order of magnitude
Overhead costs
Preliminary estimates
Quantity takeoff
Unit pricing

Review Questions

1. How much will four flights of stairs cost (including overhead and profit) if they are specified to be spiral steel stock units, primed with a 5' diameter? What crew will do this work? How long will it take to build the four flights? (Hint: see Figure 8.2)
2. A project completed in 2009 cost $3,000,000. How much would a similar project cost in 2013 using the index values provided in the table shown?

Problem 2 Index Data

YEAR	INDEX VALUE
2009	2423
2010	2697
2011	2842
2012	3100
2013	3150
2014	3167

3. What is the standard crew to install a 25" × 22" single bottom sink? How long will it take to install eight sinks? How much will it cost including overhead and profit? (Hint: see Figure 8.4)
4. What is the difference between a preliminary and detailed estimate?
5. What is the purpose of the engineer's estimate? How does the owner use it? Who prepares it? When is it prepared in the project life cycle?
6. How is a software program such as HeavyBid more flexible than use of a reference manual?

Management Pro

MANAGEMENT PRO

What factors must be considered when a contractor determines how much to mark up a project? Write a short report describing the effect of factors such as competition, type of project, economic conditions, and so forth on determining how much a contractor can mark up a bid. Can you find any mathematical models that are used to calculate markup?

References

Flyvbjerg, Bent, Nils Bruzeliu, and Werner Rothengatter. 2003. *Megaprojects and Risk: An Anatomy of Ambition*. Cambridge, UK: Cambridge University Press.

Gould, Frederick E. and Nancy E. Joyce. 2002. *Construction Project Management*. Upper Saddle River, NJ: Prentice-Hall.

Grogan, T. 2009. What Drives ENR Cost Indexes. Available from ENR.com webpage at http://enr.ecnext.com/coms2/article_bmfi090318ENRCostIndex (accessed July 25, 2009).

International Alliance for Interoperability. Technical-industry foundation classes. Available from http://www.buildingsmart.com/bim (accessed July 25, 2009).

Jackson, Barbara J. 2004. *Construction Management Jump Start*. Alameda, CA: Sybex.

Love, P. 2002. Influence of project type and procurement method on rework costs in building construction projects. *Journal of Construction Engineering and Management* 128: 18–29.

Management Computer Controls, Inc. 2007. Estimating wizards. Available from http://www.mc2-ice.com/products/iceprodsubs/wizards.html (accessed June 12, 2009).

Manfredonia, B. and J.P. Majewski. 2008. Cost estimating. Washington, DC: WBDG, National Institute of Building Sciences. Available from http://www.wbdg.org/design/dd_costest.php (accessed July 23, 2008).

On Center Software. 2005. On-Screen Takeoff. Available from http://www.oncenter.com/products/ost.asp (accessed September 26, 2005).

Timberline Software Corporation. 2004a. ePlan takeoff. Available from http://sagetimberlineoffice.com/include/pdfs/ePlan_takeoff.pdf (accessed September 26, 2005).

Timberline Software Corporation. 2004b. Model estimating. Available from http://sagetimberlineoffice.com/include/pdfs/model_estimating.pdf (accessed September 26, 2005).

Timberline Software Corporation. 2004c. Scheduling integrator. Available from http://sagetimberlineoffice.com/include/pdfs/scheduling_integrator.pdf (accessed September 26, 2005).

chapter 9

Construction Project Cash Flow and Cost Control

Chapter Outline

Introduction

After the construction contract is awarded, the contractor must manage the project during the construction phase. Two important issues for the construction contractor are obtaining the money to construct the project and monitoring costs during construction to insure profitability. In this chapter, issues related to cash flow and the need for some contractors to obtain bank loans to pay expenses during construction are discussed. Additionally, the methods and computer systems that contractors use to monitor costs during construction are illustrated.

The Cash Flow Problem

Now that we have studied the contractual relationships that occur on a construction project, it is useful to consider how construction contractors are paid, and how it affects their behavior during a construction project. Chapter 2 discussed how the contractual relationship between the owner and the contractor are described in the general conditions. Importantly, the mechanism for the payment of the construction contractor during construction is described in the general conditions.

For most construction projects, the general conditions state that a contractor shall submit a request for payment on a monthly basis. For a unit price contract, the amount paid represents the actual installations made during the pay period. For lump sum projects typical in building construction, the amount paid is on the basis of an estimate of project progress made by the designer. The contractor submits this request for payment to the owner and usually receives the payment after a time lag of several weeks. The request for payment is reviewed and approved by the designer (an engineer or architect depending on the type of construction).

With this system of payments, it can be seen that payments due to a construction contractor generally lag behind the contractor's expenditures on a project. This introduces a **cash flow problem** for the contractor. A company's **cash flow** can be defined as the pattern of a company's income and expenditures and the resulting availability of cash. Construction contractors must make sure they have enough money on hand to build the project; yet their payments lag a month or more behind their expenditures. Expenditures on a construction project tend to take the form of a continuous S-shaped curve. It indicates that projects start slowly, quicken during the middle phases, and then wind down.

Often, construction contractors do not have enough cash on hand to make up for the difference between their expenses and their revenues. To have enough cash to complete the project, the construction contractor must get a loan from the bank called a **line of credit**. With a line of credit, interest begins to accrue immediately as funds are drawn from the bank. A contractor arranges the line of credit to insure that he or she will have enough cash available to build the project. Lines of credit are issued by commercial banks. The line of credit is relatively easy to obtain but the amount of the line of credit must be arranged with the bank before beginning the project. A contractor must establish his or her spending limit with the bank before the project commences.

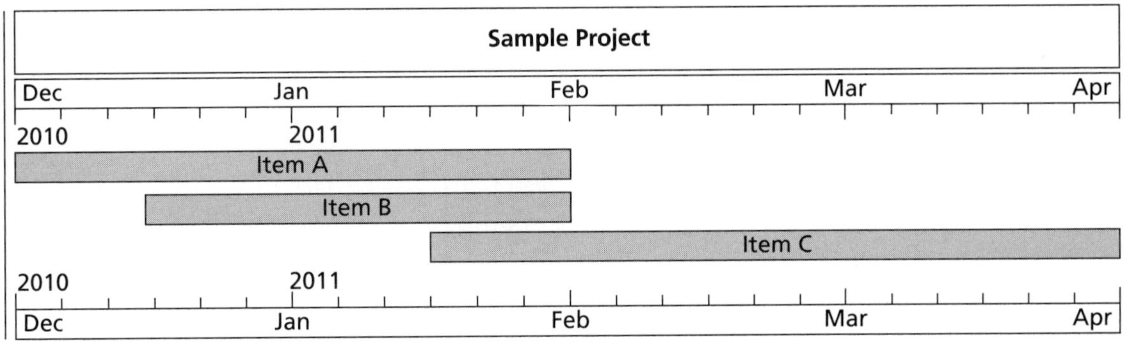

Figure 9.1 Example project bar chart

Contractors negotiate with the bank to determine the repayment terms. Some types of loans require only monthly interest payments, which can sometimes be advantageous for small construction contractors. A study of electrical contractors found that highly profitable companies with an annual direct labor payroll of $1 million or less had an average line of credit of $335,000 (Jaselskis et al., 2002).

A Cash Flow Example

Figure 9.1 shows a bar chart for a three-activity project. The project is scheduled to take 4 months to complete. Table 9.1 shows a contractor's worksheet used to calculate the bid prices for this unit price project. Column 3 shows the unit prices the contractor has calculated to cover his or her costs. Column 4 shows the contractors total cost to complete each line item (Column 2 × Column 3). Column 5 shows the contractor's bid price, which has been marked up by 8% to provide the contractor with a profit. (The contractor decides the percentage markup to use.) The total cost plus profit column is the contractor's bid price. The cost plus profit for an item is given as:

$$\text{Cost plus profit} = \text{Total cost} \times 1.08$$

Table 9.2 illustrates the additional calculations necessary to determine the amount of the overdraft loan that will be needed to cover the contractor's expenses during the construction period. The owner deducts a 10% retainage during the project. Assume that work effort is linear across the life of each project item. Assume also, in Month 1

Table 9.1 Unit prices and item cost including profits

ITEM (1)	QUANTITY (UNITS) (2)	UNIT PRICE ($/UNIT) (3)	TOTAL COST (4)	COST PLUS PROFIT (5)
A	1,000	25	25,000	27,000
B	100	120	12,000	12,960
C	1,000	25	25,000	27,000
Totals			62,000	66,960

Table 9.2 Calculation of overdraft

MONTH (1)	MONTHLY EXPENSES (2)	CUMU-LATIVE EXPENSES (3)	AMOUNT BILLED (4)	CUMU-LATIVE BILLINGS (5)	RETAIN-AGE (6)	PAYMENT RECEIVED (7)	CUMULA-TIVE PAY-MENTS RECEIVED (8)	OVER-DRAFT BEFORE INTEREST (9)	INTEREST (10)	OVER-DRAFT AT END OF MONTH (11)	OVER-DRAFT AFTER PAYMENT RECEIVED (12)
1	16,460	16,460	17,777	17,777	1,778	0	0	16,460	165	16,625	16,625
2	25,540	42,000	27,583	45,360	2,758	15,999	15,999	42,165	422	42,586	26,587
3	10,000	52,000	10,800	56,160	1,080	24,825	40,824	36,587	366	36,953	12,128
4	10,000	62,000	10,800	66,960	1,000	9,720	50,544	22,128	221	22,349	12,629
5	0	62,000	0	66,960	0	16,416	66,960	12,629	0	12,629	−3,787

of the project, that the contractor expects to expend 50% of the estimated cost for Item A and 33% of the cost for item B. To calculate this amount, multiply the total cost (Column 4 of Table 9.1) of each scheduled project item by the percentage of each item that will be expended in that month. Therefore, the contractor's expenses in the first month of the project are:

$$25{,}000 \times 0.5 + 12{,}000 \times 0.33 = 16{,}460$$

The amount of money billed to the owner in Month 1 must include the profit. The owner will be billed using the cost plus profit amount (Column 4 in Table 9.2):

$$27{,}000 \times 0.5 + 12{,}960 \times 0.33 = 17{,}777$$

This would be the amount the contractor would bill the owner at the end of Month 1. The amount billed for each month is shown in Column 4 of Table 9.2. In this project, the owner pays the contractor 30 days after the monthly bill is submitted.

The gap between expense outflows and monthly payments is further widened by the retainage deducted from the monthly progress payment by the owner. Typically, the owner retains 5–10% of the amount owed to the contractor as retainage. These funds are paid to the contractor only upon substantial completion of the project. In this project, as we know, the owner charges a 10% retainage. Column 2 of Table 9.2 shows the expenses incurred by the contractor each month, and Column 3 shows the cumulative expenses. The retainage is calculated as:

$$\text{Retainage} = \text{Amount billed} \times 0.1$$

The retainage deducted for each month of the project is shown in Column 6 of Table 9.2. Column 7 shows the payment received from the owner and is the contractor's revenue for the project. Payment is received 30 days after the contractor submits the invoice to the owner. For example, the payment received in Month 2 is given by:

$$\text{Payment received (Month 2)} = \text{Amount billed (Month 1)} - \text{Retainage (Month 1)}$$

Figure 9.2 shows a plot of the contractor's cumulative expenses and cumulative revenues. The difference between the two curves at any time represents the amount of overdraft required by the contractor. The figure illustrates how the payments lag behind the contractor's expenditures.

A large enough line of credit must be arranged at the bank before the initiation of the project. The contractor can project his or her loan requirements by performing some additional calculations. The overdraft at the end of the first month of the project will be the first month's expenses multiplied by the interest rate. For this problem, we assume a monthly interest rate of 1%. Table 9.2 shows the calculation of the end-of-month overdraft in two stages. Column 9 shows the amount of overdraft at the end of the month before interest is added. It is given by:

$$\begin{aligned}\text{Overdraft before interest (Month } N) = {} & \text{Overdraft after payment} \\ & \text{received (Month } N - 1) \\ & + \text{Monthly expenses (Month } N)\end{aligned}$$

Figure 9.2

Project income and expense curves

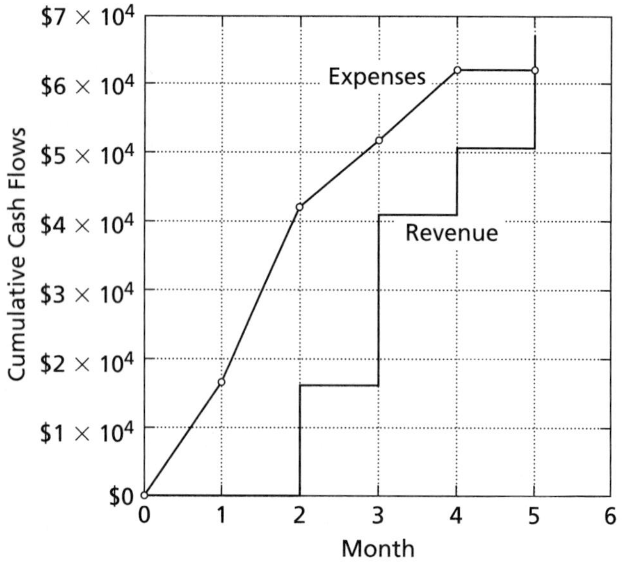

The overdraft at the end of the month is equal to the overdraft before interest multiplied by the interest rate of the overdraft loan. The overdraft loan required at the end of the month is shown in Column 11 of Table 9.2. In this problem, we have assumed a linear expenditure of expenses. Therefore, the maximum overdraft level within a month will occur at the end of the month just before the payment from the owner arrives and is the amount shown Column 11. The overdraft is reduced when the monthly payment arrives and is shown in Column 12. The overdraft after payment received is given as:

$$\text{Overdraft after payment received (Month } N) = \text{Overdraft at end of month (Month } N) - \text{Payment received (Month } N)$$

Examination of Table 9.2 indicates that a negative number is shown for the overdraft after payment in the last week of the project. This is actually the profit on the project less the amount of interest paid on the overdraft loan.

Figure 9.3 shows a plot of the maximum overdraft and the overdraft after payment is received for each month of the project. The plot shows that the maximum overdraft for the whole project will occur in Month 2. To have the cash to build the project, the contractor must have arranged a line of credit at the bank larger than $42,586.

Unbalanced Bids

Contractors bidding on unit price contracts can unbalance their bids to reduce the amount of the overdraft. To unbalance a bid, the contractor first calculates his or her unit prices for all line items on the proposal form, as well as summing these items to determine the total project cost. Then the contractor increases the unit price for some items early in the project and decreases the unit prices for some items late in the project. The contractor thus maintains the same total project cost as originally calculated so the bid remains competitive. Although this is common practice in construction, the

Figure 9.3

Monthly overdraft

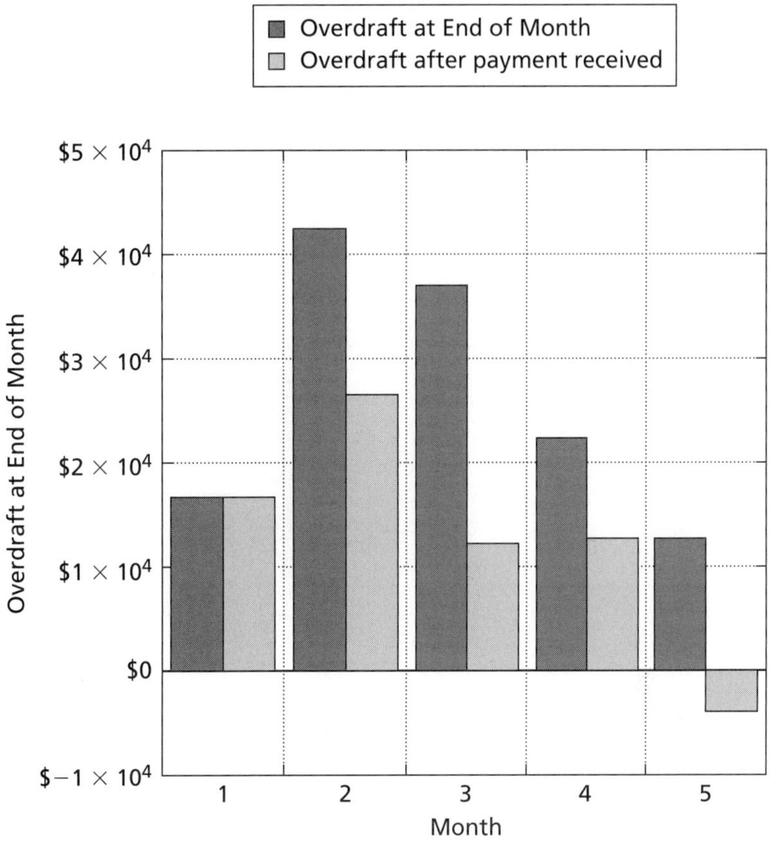

unbalancing cannot be too blatant because an owner might reject the bid as being not responsible at the bid opening.

Table 9.3 shows the sample problem with the unit prices readjusted to earn more revenue in the early stages of the project. The unit price for the first item has been increased and the third item has been decreased. Table 9.4 shows the revised overdraft calculations (for the unbalanced situation). The total project cost has been maintained. Figure 9.4 shows a plot of the expense and payment curves for this project with unbalanced unit prices. Figure 9.5 shows that the overdraft requirements for Months 3–5

Table 9.3 Unbalanced unit prices and item cost including profits

ITEM	QUANTITY (UNITS)	UNIT PRICE ($/UNIT)	TOTAL COST	COST PLUS PROFIT
A	1,000	44	44,000	47,520
B	100	80	8,000	8,640
C	1,000	10	10,000	10,800
Totals			62,000	66,960

Table 9.4 Overdraft calculation with unbalanced unit prices

MONTH	ACTUAL EXPENSES	CUMULATIVE EXPENSES	AMOUNT BILLED	CUMULATIVE BILLINGS	RETAINAGE	PAYMENT RECEIVED	CUMULATIVE PAYMENTS RECEIVED	OVERDRAFT BEFORE INTEREST	INTEREST	OVERDRAFT AT END OF MONTH	OVERDRAFT AFTER PAYMENT RECEIVED
1	16,460	16,460	26,611	26,611	2,661	0	0	16,460	165	16,625	16,625
2	25,540	42,000	31,709	58,320	3,171	23,950	23,950	42,165	422	42,586	18,636
3	10,000	52,000	4,320	62,640	432	28,538	52,488	28,636	286	28,923	385
4	10,000	62,000	4,320	66,960	1,000	3,888	56,376	10,385	104	10,488	6,600
5	0	62,000	0	66,960	0	10,584	66,960	6,600	0	6,600	−3,984

Figure 9.4

Project income and expense curves with unbalanced bid

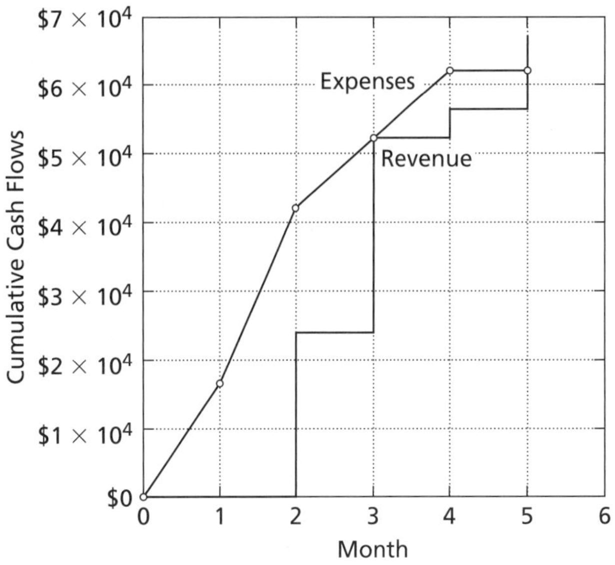

Figure 9.5

Monthly overdraft with unbalanced bid

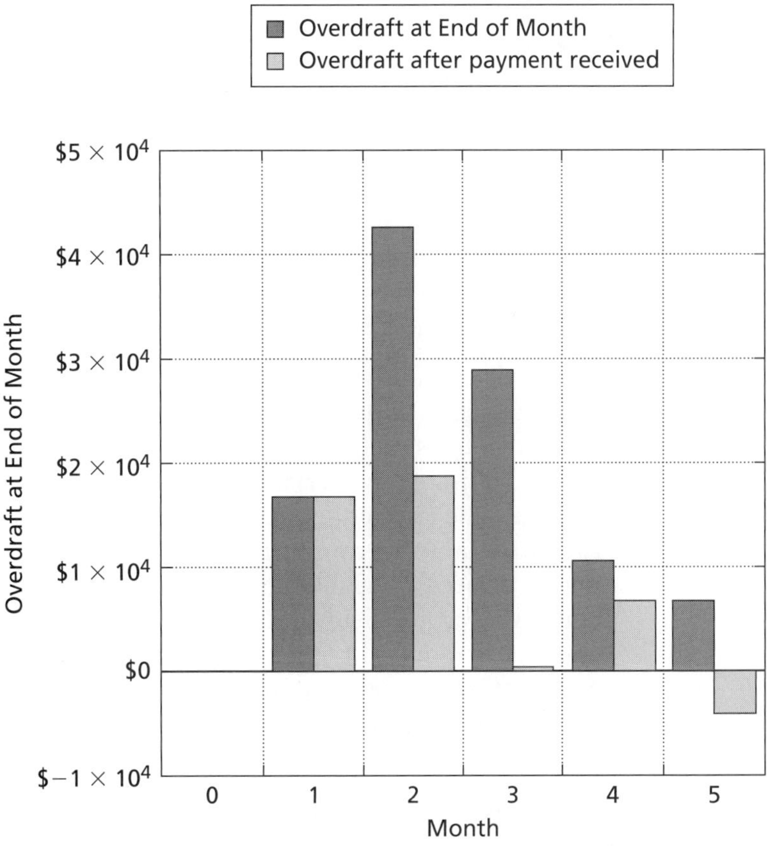

are significantly reduced. However, the maximum overdraft is still the same with the **unbalanced bid** because the first payment is still not received until the end of Month 2, the month where the maximum overdraft occurs.

Mobilization Payments

Mobilization is often used by owner organizations as a means of reducing the overdraft loan. Construction contractors are in a risky business, and banks often charge contractors a high rate of interest for overdraft loans. Because contractors include the interest cost of the overdraft loan in their bid price, an owner can reduce the cost of overdraft loan interest by paying a mobilization fee at the initiation of the project. Large owners either have the cash or can borrow at lower interest rates than the contractor. Therefore, the entire project cost is reduced by the difference between the contractor's borrowing cost and the owner's borrowing cost.

Project Cost Control

In the first part of this chapter, we studied how a contractor makes sure he or she has enough money to complete the construction project as payments lag behind expenses. This section of the chapter addresses the related issue of controlling project costs. The purpose of all construction companies is to make a profit. A contractor must have a cost control system in place to monitor expenditures to insure that costs do not increase above the original estimate amount. Additionally, a cost control system can be a valuable source of information for estimating future projects. The cost control system accumulates the construction company's actual costs on all the projects it conducts. A construction project will have many items that generate expenses for the contractor. The contractor must pay for the materials to construct the project, the cost of equipment, and the cost of construction labor used on the project. Overhead costs must also be considered when calculating project costs.

Materials are all the items that are needed to construct a project. To construct a project, thousands of purchases will need to be made by the contractor. These vary from mundane items such as nails to complex heating and air conditioning units for a commercial building. Material purchases are accomplished through the use of purchase orders, which are collected in the cost control system. Labor costs are the costs of the direct labor to construct the project. This includes the cost of construction labor on the project. The labor cost data are collected from the time sheets filled out by the employees in the field. Equipment costs are the costs associated with use of heavy equipment required for the construction project. For example, if a crane is rented for a project, the crane generates costs as long as it is in use on the project. In a project where subcontractors are used, the general contractor must monitor their costs also to insure that their progress corresponds to their billings.

Overhead is the cost of equipment, materials, and services that are necessary for the construction company to conduct business but are not related directly to the construction work. As indicated in Chapter 8, home office overheads are the costs related to maintaining an office with managers, estimators, accountants, marketing specialists, and clerical support staff that are required to manage the firm but are not directly related to the physical construction of a facility. Project overheads are the costs associated with supervising the construction of a specific project. This includes the cost of supervisory personnel such as project managers, and superintendents, and the cost of temporary project offices. These overhead costs are included in the project's bid prices and must also be tracked during a project to make sure the funds budgeted for these support activities are not exceeded.

For any construction cost control system, the starting point is the original project estimate. This becomes the budget for the construction project. As costs are incurred during construction, the costs will be compared with the original budget. When costs are observed to exceed the original budgeted amount, it is an indication of problems in the field that must be addressed.

Computerized Cost Control Systems

Good construction companies apply a systematized approach to controlling their costs. Most companies use computerized systems to monitor costs. The data collected from the field are entered into a computerized cost control system. For a computerized system to work, it is necessary to develop standardized cost codes so the data can be retrieved and used. Typically, a firm will develop a set of standard cost accounts that will vary depending on the type of work that a firm does. The complexity of a coding scheme can vary depending on the nature of the work and the information the contractor requires to manage the project. Additional information that can be included in a code is the location of the work and the associated "work package."

Example

For cost estimating and scheduling purposes a project can be broken into hierarchical elements. This is called a work breakdown structure. A work package is an element within the work break down structure. Therefore, a cost coding scheme could be used that identifies project costs with the various work packages that make up the hierarchy of project costs.

In Chapter 2, the Construction Specifications Institute (CSI) standard cost codes were discussed as a way of categorizing project specifications for commercial buildings. The use of these codes can be extended to the cost control system, where the data collected from the field are stored and classified according to the CSI code it is associated with. The same codes used in the specification documents can also be used to identify project costs.

JobView is a construction accounting and cost control software. It is intended to be used by small- and medium-sized contractors. It provides a good illustration of the inputs and outputs required for a cost control system. Figure 9.6 shows a purchase order

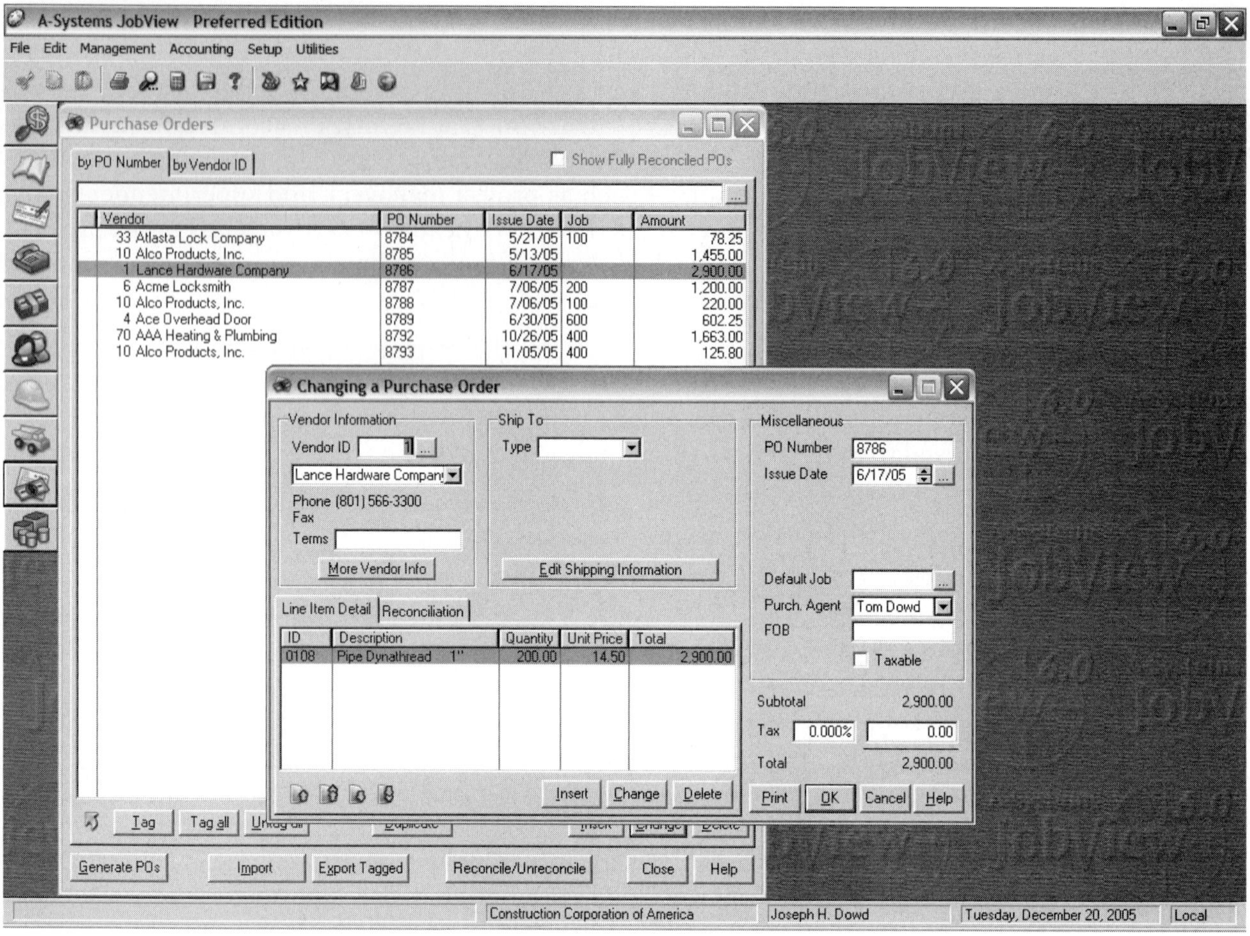

Figure 9.6 Purchase order input to computerized cost control system. *Used courtesy of A-systems JobView®.*

input form where material purchases are recorded in the cost control system. Figure 9.7 shows a form that is used to input information on payments to subcontractors. The window in the background lists the status of payments to all the subcontractors working on a project.

Figure 9.8 shows an important table that is generated by the JobView software. This table shows all project items listed by their cost code. It allows for any analysis of project costs to date. Notice that the figure shows three budget numbers: original, current, and remaining budgets. The original budget is from the project estimate. The current budget is the amount of money allocated to a cost code at a given point in the project. This number may vary because of change orders, with more funds allocated to the cost code. The remaining budget is the amount of money left to be spent for the cost code. The percentage complete of any cost code is given by the actual cost to date divided by the projected cost. The current variance is the difference between the current budget and actual costs. A positive number in the variance column indicates a cost overrun and a negative number indicates

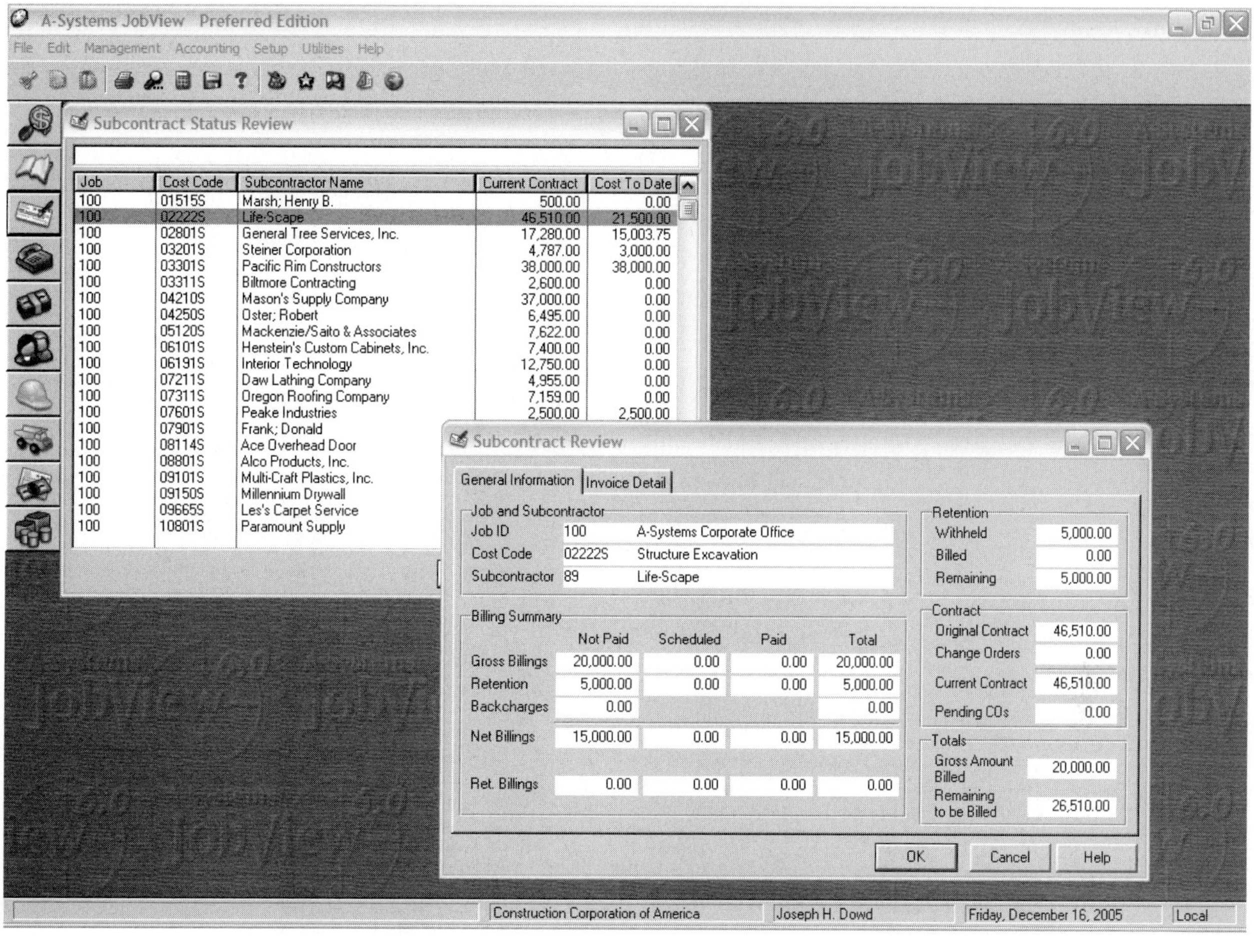

Figure 9.7 Monitoring subcontractors using cost control system. *Used courtesy of A-systems JobView®.*

a situation where the product has been constructed for less than the estimate. The projected cost is the actual cost to date plus the estimated cost to complete the work for the cost code. A projected overrun is the difference between the projected cost and the current budget. Again, a positive number indicates that budgeted costs have been exceeded.

The type of information shown in Figure 9.8 is commonly used by construction contractors. It allows for the identification of work items that are deviating from the budget. For example, the cost code 03301 G, pour footings, has exceeded budgeted costs. This report provides a "red flag" that management's attention must focus on this work to identify what has happened and complete construction while minimizing expenses. Cost overruns indicate potential inefficiencies are occurring that must be fixed to prevent additional overruns. The cost overrun may also have been the result of an estimating error, in which case the information from the report allows estimators to consider revising future estimating methods to produce more accurate budgets.

In analyzing Figure 9.8, it is important to understand the limitations of this type of cost analysis. The percentage completion of a cost item is measured by the amount

Construction Corporation of America
Complete Job Cost Analysis

12/16/05
10:16AM

Job 100

Cost Code	Description	Quantity Estimate	Quantity To-Date		% Complete Report Calc	Budget Original	Budget Current	Remaining	Current Variance	Actual Cost To-Date	Projected Cost	Projected Overrun
100	**A-Systems Corporate Office**				Contract	600 000.00	600 000.00					
01000 G	General Requirements		0.0		0	10 109.00	10 109.00	10 109.00	0.00	0.00	10 109.00	0.00
01000 L	General Requirements	23.0	69.0	Hours	100	0.00	0.00	-826.73	826.73	826.73	826.73	826.73
01011 L	Supervision	258.0	400.0	Hours	100	8 500.00	8 500.00	4 106.00	-4 106.00	4 394.00	4 394.00	-4 106.00
01020 G	Overhead		0.0		0	25 777.00	25 777.00	25 777.00	0.00	0.00	25 777.00	0.00
01030 G	Profit as Bid		0.0		0	26 479.00	26 479.00	26 479.00	0.00	0.00	26 479.00	0.00
01515 S	Project Signs	Marsh; Henry B.			0	500.00	500.00	500.00	0.00	0.00	500.00	0.00
01607 G	Building Survey		0.0		0	1 500.00	1 500.00	1 500.00	0.00	0.00	1 500.00	0.00
01609 G	Subsistence		0.0		0	60 630.00	60 630.00	60 630.00	0.00	0.00	60 630.00	0.00
General Requirements	Subtotal:	281.0	469.0	Hours	4	133 495.00	133 495.00	128 274.27	-3 279.27	5 220.73	130 215.73	-3 279.27
02222 S	Structure Excavation	Life-Scape			46	46 540.00	46 510.00	25 010.00	0.00	21 500.00	46 510.00	0.00
02801 S	Landscaping	General Tree Services, Inc.			87	17 280.00	17 280.00	2 276.25	0.00	15 003.75	17 280.00	0.00
Sitework	Subtotal:		0.0	Hours	57	63 820.00	63 790.00	27 286.25	0.00	36 503.75	63 790.00	0.00
03201 S	Rebar Reinforcing	Steiner Corporation			63	4 787.00	4 787.00	1 787.00	0.00	3 000.00	4 787.00	0.00
03301 G	Pour Footings		0.0		100	1 000.00	1 000.00	-542.69	542.69	1 542.69	1 542.69	542.69
03301 L	Pour Footings		16.0	Hours	15	1 000.00	1 000.00	854.78	0.00	145.22	1 000.00	0.00
03301 M	Pour Footings		0.0		0	40 000.00	40 000.00	40 000.00	0.00	0.00	40 000.00	0.00
03301 S	Pour Footings	Pacific Rim Constructors			100	38 000.00	38 000.00	0.00	0.00	38 000.00	38 000.00	0.00
03311 G	Pour Ground Slabs		0.0		0	1 600.00	1 600.00	1 600.00	0.00	0.00	1 600.00	0.00
03311 L	Pour Ground Slabs	33.0	0.0	Hours	0	1 000.00	1 000.00	1 000.00	0.00	0.00	1 000.00	0.00
03311 M	Pour Ground Slabs		0.0		0	5 700.00	5 700.00	5 700.00	0.00	0.00	5 700.00	0.00
03311 S	Pour Ground Slabs	Biltmore Contracting			0	2 600.00	2 600.00	2 600.00	0.00	0.00	2 600.00	0.00
Concrete	Subtotal:	33.0	16.0	Hours	44	95 687.00	95 687.00	52 999.09	542.69	42 687.91	96 229.69	542.69
04210 S	Brick Masonry	Mason's Supply Company			0	37 250.00	37 000.00	37 000.00	0.00	0.00	37 000.00	0.00
04250 S	Ceramic Veneer	Oster; Robert			0	6 495.00	6 495.00	6 495.00	0.00	0.00	6 495.00	0.00

Figure 9.8 Job cost analysis generated by JobView software. *Used courtesy of A-systems JobView®.*

of money spent divided by the budget. However, the amount of money spent may not be indicative of actual progress in constructing the cost item. In Chapter 10, on scheduling, we will discuss how the cost control system can be linked to the schedule to provide a more meaningful analysis of project costs.

Summary

This chapter has focused on two major issues that construction companies must address. The first is the problem of cash flow and the arrangements a contractor must make to insure that he or she has enough money to conduct the project work while waiting for payments from the owner to arrive. The second issue is the contractor's need to control costs during construction to insure that the project is profitable. Construction contractors typically maintain computerized cost control systems that allow project costs of current projects to be tracked and provide a database of construction cost information for estimation in future projects.

Key terms

- Cash flow
- Cash flow problem
- Line of credit
- Mobilization
- Unbalanced bid

Review Questions

1. Why is cash flow a problem for construction contractors?
2. Discuss how an unbalanced bid can reduce the overdraft loan requirement.
3. What is the purpose of mobilization? When is a mobilization payment made?
4. What is the purpose of a cost control system?
5. In Figure 9.8, identify the cost codes that are currently over budget.

Management Pro

MANAGEMENT PRO

A three-activity construction project has a schedule as in the figure shown. The contractor's profit will be an 8% markup of all costs. The table shows the contractor's costs for each project work item. The contractor has negotiated a line of credit at the bank that will have a 1% monthly interest rate. Retainage on the project will be 5%. The contractor will submit invoices for payment at the end of each month and payment from the owner will be received 30 days later. What is the maximum overdraft for this project? During what month of the project will it occur?

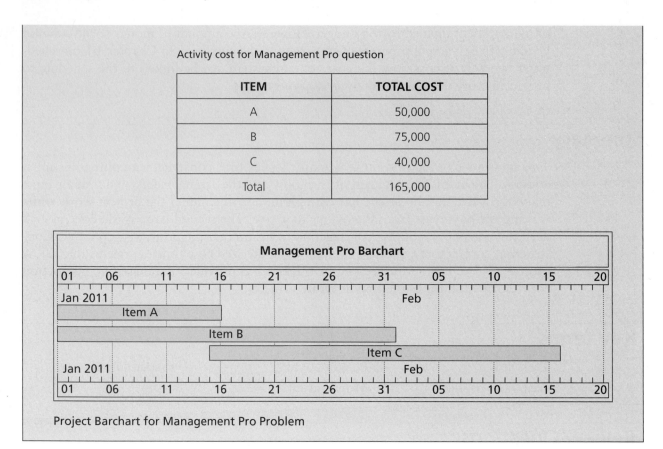

Activity cost for Management Pro question

ITEM	TOTAL COST
A	50,000
B	75,000
C	40,000
Total	165,000

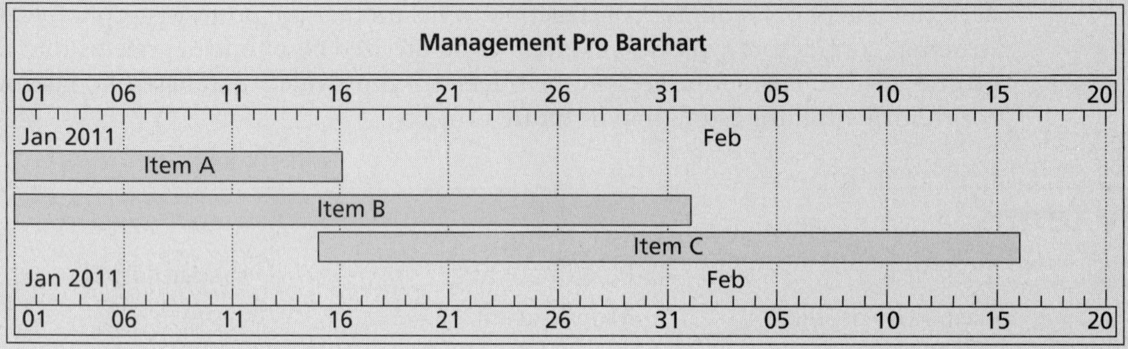

Project Barchart for Management Pro Problem

Reference

Jaselskis, Edward J., James M. Kurtenbach, and John Forrest. 2002. Enhancing financial success among electrical contractors. *Journal of Construction Engineering and Management* 128 (1): 65–75.

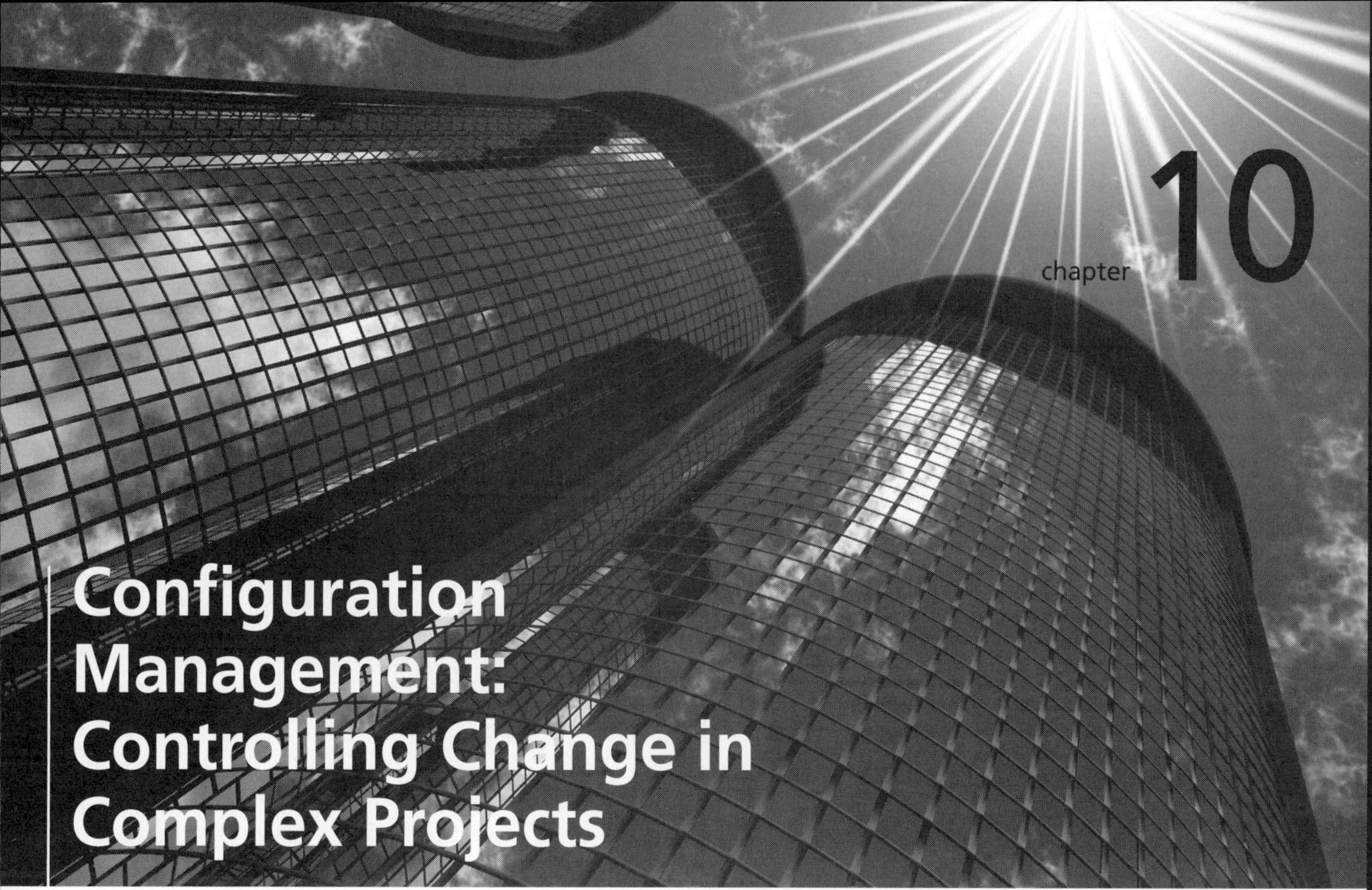

chapter 10

Configuration Management: Controlling Change in Complex Projects

Chapter Outline

Introduction

Change is an important part of the construction of a large project. All through the project cycle, changes will be made to the project. Starting with the project planning process, various alternatives for the project will be considered and an initial project scope developed. Preliminary cost estimates will be compared with the available funding. During design, changes will occur as the designer and owner interact to develop detailed designs and specifications. As construction proceeds, changes will occur that may have significant impacts as constructability issues arise and errors or omissions are found in the design. All these changes must be carefully managed to insure that project costs and schedule remain in control. Often, complex projects can experience "scope creep" if changes are not well controlled. **Scope creep** occurs when many seemingly small changes accumulate over time and change the scope of the project into something larger and costlier than what was originally intended. Configuration management is a technique that is being increasingly used in the construction industry to manage the change that occurs in complex projects to help keep projects from experiencing scope creep, large cost overruns, and major schedule slippages.

Configuration Management

Configuration management (CM) has been used since the 1950s; however, it has only recently gained acceptance as a project management tool in the construction industry. Configuration management can be defined in different ways depending on the industry and the type of application. The American National Standards Institute defines CM as "a management process for establishing and maintaining consistency of a product's performance, functional, and physical attributes with its requirements, design, and operational information throughout its life" (American National Standards Institute, 1998). In a construction context, the "product" is the constructed facility that is the outcome of a construction project.

Configuration management principles were first documented and adopted by the United States Department of Defense. The military's focus on CM started when concerns over life safety issues were raised during the design and manufacture of weapons and aircraft. Subsequently, the nuclear industry adopted CM to track equipment changes affecting operational safety, and today CM is actively used in many industries, including the software development industry to manage version control as complex computer software are modified and developed.

An understanding of configuration management requires that "configuration" be defined. For a construction project, a configuration can be defined as the functional and physical characteristics of the project as set forth in technical documentation (plans and specifications) ultimately achieved as the completed construction (Department of Defense Systems Management College, 2001). The configuration is what is to be built. If the plans and specifications define what is to be built, any change to these documents, such as a change order, is a change in the configuration. In complex construction projects, this configuration is constantly being modified throughout the entire life cycle of a project, from the conception of the project through its completion and use.

A complex project goes through many stages. At all the stages, there are many decisions that are made that affect the configuration of the project. Consider the development of a subway system and the many phases of the project where the configuration is modified:

- **Planning.** A government agency conceives of the need for a subway system and defines its objectives. In the planning phase, various routes and alignments must be considered. The routing will change over time as more information is gathered about constraints to various alignments. Preliminary estimates will rule out some alternatives as too costly. The configuration of the system will be modified often as the system alignment is developed.
- **Design.** Discussions between the owner, engineers, and architects will affect the configuration as the design increases in detail. The design will be complex, including civil design of the track, elevated structures, and rail. Architects will design the stations and mechanical and electrical engineers will need to design the subway cars and the required electrical control systems and signals. Many alterations to the project configuration are possible.
- **Construction.** Change orders will occur because of errors and omissions in the plans or because of constructability issues. Again, this will cause further modification to the project configuration.
- **Completion and operation.** To operate the system correctly, the owner requires documentation on the configuration of the completed facility. The owner must verify that the constructed facility is properly configured.

Clearly, a major project like a subway system will have thousands of changes to its configuration during the course of a project. In complex projects, the number of changes that occur will be beyond the understanding of a team of managers. Therefore, the discipline of a formal configuration management system is required to control change in a rational manner. In construction, configuration management is the careful management of changes to the configuration of the project to insure that the completed project meets the owner's objectives and requirements.

Case Study: The Benefits of Employing Configuration Management—The Los Angeles County MTA

The Los Angeles County Metropolitan Transportation Authority was responsible for building the Red Line subway in downtown Los Angeles. After experiencing significant budget overruns in the first segment of the system, a configuration management system was implemented to better control changes in the second segment. Table 10.1 provides a comparison of costs, change orders, and claims for the two segments (PACO Technologies, 2007c) and indicates clearly that the second

Table 10.1 Comparison of costs, changes, and claims in Los Angeles Red Line Segments 1 and 2

	RED LINE SEGMENT 1	RED LINE SEGMENT 2	PERCENTAGE CHANGE
Base value of awarded contracts	$664 million	$1,048 million	+58
Number of logged contractor claims	2,165	607	−71
Estimated value of logged claims	$87 million	$46 million	−47
Total number of logged changes and claims	4,728	4,023	−14
Total estimated value of changes and claims	$153 million	$121 million	−20
Estimated increase over award value	24%	12%	−50

segment was better managed. The Los Angeles MTA reported several benefits of implementing the configuration management system:

- A 60% reduction in administrative costs, resulting in $10 million in savings
- Reduced time to process change orders, resulting in a reduced number of contract claims
- Reduced response time to information requests, from days to minutes
- Increase in time available for managers to spend on analysis and management of project issues

The experience of the Los Angeles MTA quantifies the potential benefits to be gained from application of configuration management techniques. Although there are costs associated with the implementation of configuration management, including the purchase of a document management system and the costs of enforcing a formalized system, they are small compared with the potentially large benefits of better project control (Steinberg and Otero, 2007).

Claims Management: Another Benefit of Configuration Management

The Tren Urbano is a mass transit system that has recently been constructed in San Juan, Puerto Rico. The Tren Urbano is composed of 16 stations, 10 of which are elevated, 4 at grade or in open cuttings, and 2 underground. A maintenance depot and operations control center is provided halfway along the route. Each of the 16 stations boasts unique artwork and architectural style. The project was built by letting six separate design-build contracts. A separate contract for the signal, traction power, and vehicles was also let. Each of these contracts included heavy civil construction and the construction of station buildings.

Configuration management was not initially applied to this project. It was employed in the later stages of the project as a way of better managing project documents and to provide an improved audit trail of change order work. This involved physically scanning several million documents. The configuration management system is now proving useful to the owner as a way of documenting project decisions and events to counter claims from the various design-build constructors. The configuration management system provides an audit trail of documents that gives an overview of how and when decisions were made on the project. On the Tren Urbano project, a configuration management IT system provided document control to produce a cross-referenced and searchable database of project documentation. The configuration management software allowed a scanned copy of the actual document to be viewed. A key to configuration management document control on the Tren Urbano project was the document numbering process. Each official project document was given a unique number before it was scanned. The project engineer's staff initiated this numbering for each of the separate contracts.

Implementing a Construction Configuration Management System

Implementation of a CM system involves several interrelated activities that insure that changes to a project that occur throughout the project cycle occur in a rational manner and provide a delivered facility that meets the owner's needs. These include:

- **The development of a configuration management plan that defines implementation responsibilities and procedures.** The plan defines who is responsible for initiating and approving changes to the project. The plan also defines the paperwork procedures that will be used during the project.
- **The establishment of a baseline of the original project configuration.** This allows all proposed changes to the project to be measured against the original defined goals of the project.
- **A document management system.** A formal management system is established to provide a database that archives all paperwork related to changes in the project configuration.
- **A formal system that examines proposed changes to the construction project.** In addition to insuring that changes are in line with project scope and goals, a CM system examines each change order's impact on the project. Thus, it goes beyond merely tracking change order approval documents, as some construction Web portals do.
- **A system to document "lessons learned" from the project.** CM applied to construction allows for the auditing of the performance of the completed project

and the ability to extract lessons learned from the project. A common problem in construction is the repetition or errors from project to project; application of CM allows an owner to become more knowledgeable.

The Configuration Management Plan

The effective use of configuration management in a construction project requires that a configuration management plan be developed. Central to configuration management is the collection and archiving of documents related to changes in the project configuration. A configuration management plan identifies the parties in the construction project (owner, designer, contractor) who will participate in the CM process and what their responsibilities will be in collecting and managing project documentation, as well as responsibilities for approving changes to the project configuration. The configuration management plan makes certain that (PACO Technologies, 2007b):

1. The key design and contract documents are identified and that a system is established to track changes to these documents throughout the project life cycle.
2. All project correspondence and submittals are processed and filed systematically. A major construction project will have thousands of submittals. The configuration management system provides a method of tracking the submittals and insuring that the documents are submitted to the correct parties for approval.
3. The plan provides a method of establishing an audit trail for each change request. An audit trail means that it is possible to retrieve all the documents and correspondence related to a request for a change order from the initiation of the request, redesign, and negotiation of the price.
4. Records of decisions are maintained and are easily retrievable. On complex, long-lived construction projects, thousands of decisions must be made by the owner, designer, and contractor. Often, decisions made several years in the past will need to be revisited as problems arise in construction. Many projects have been delayed because it has been impossible to find (missing) documents that describe the justifications or details for various elements of the design.

Document Management

Using computers greatly facilitates the application of configuration management. Document management is a type of computer application that aids in the archiving and retrieval of documents. Document management can be defined as the process of managing documents and other types of information such as images, from creation, review, storage, to their dissemination. It also involves the indexing, storage, and retrieval of documents in an organized way. A document management system also allows for hard copy documents to be scanned and stored in the system. A configuration management computer tool will be described in following text that employs powerful document management tools to archive, cross-reference, retrieve, and search project documentation.

Successful implementation of configuration management on a project requires that all documentation and correspondence are closely managed, tracked, and archived

(Steinberg and Otero, 2007). Documents that are formally managed include schedules, diagrams, plans, specifications, requests for information, change requests, and all written daily project correspondence. This level of detail is necessary to avoid changes to project documentation that has not followed the formal review procedure defined in the configuration management plan.

Establishing and Managing a Project Baseline

One of the most important aspects of applying configuration management to a construction project is the establishment of a project baseline. A **baseline** can be defined as the formal documenting of a product at some level of design definition (Department of Defense Systems Management College, 2001). For a construction project, the project baseline describes the requirements and functions of the facility that is to be constructed. The baseline for a project should be defined as early as possible in the project life cycle. In the construction industry, the original baseline will contain preliminary design documents, a functional scope, and preliminary estimates establishing the project budget and the project schedule. The project baseline and all preliminary design, estimate, and trade-off analyses are retained for future use as the project progresses. As changes occur, new baselines illustrating the project configuration will be produced. Figure 10.1 illustrates the typical documents included in the baseline, the latter's link to document management, and the completed project.

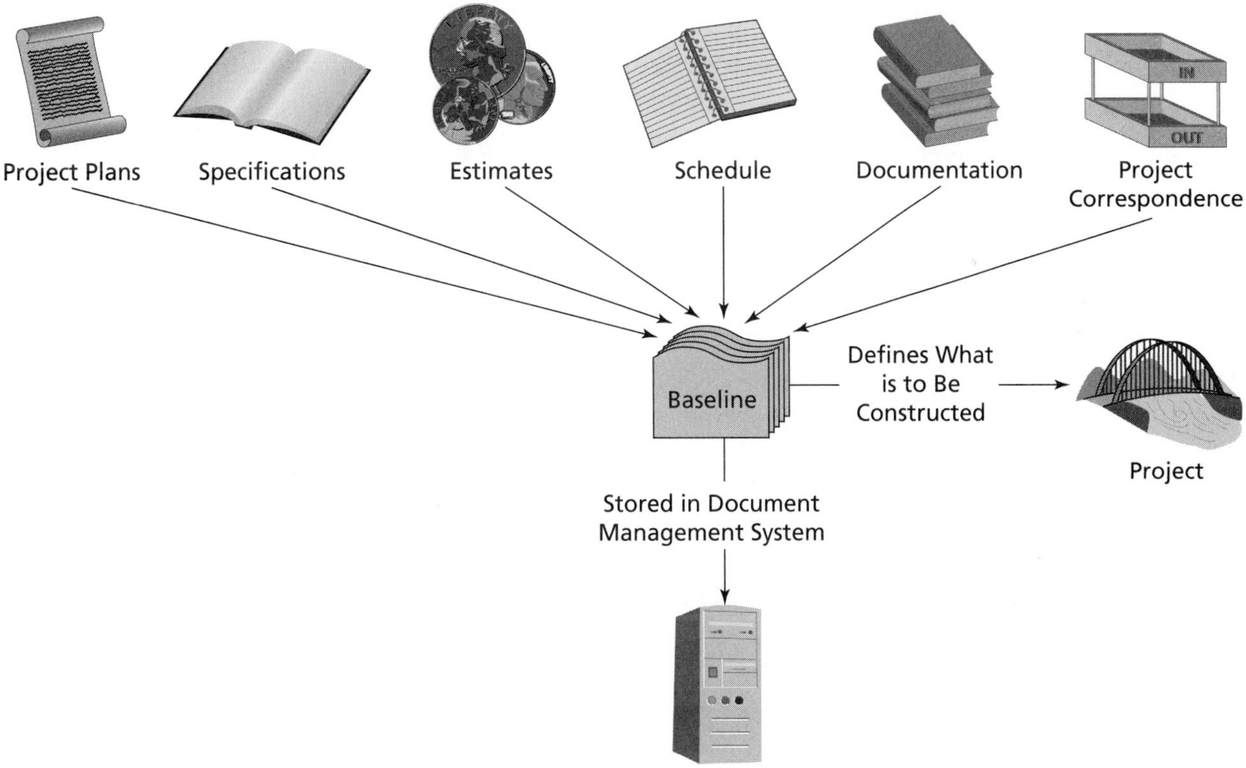

Figure 10.1 Components of a configuration management system

As the project progresses, an important component of configuration management is to track and document the changes to the project. As a construction project is modified, it is always possible to compare the existing configuration of the project with the original project baseline to determine how extensively the project has been modified from it original scope. Baseline management involves "the capture and archive of the exact state of a project during key points of a project lifecycle" (Steinberg and Otero, 2007). All documents generated during a project are archived. A key to the application of configuration management is that the documents are not only stored but also the relationship between documents is captured; that is, a letter requesting a design change is stored in such a way that all the other documents about the design change are linked to it and can be retrieved.

Change Management

Change management is the process of reviewing and approving or denying changes to the project scope, schedule, and budget (Steinberg and Otero, 2007). The formal establishment of project baselines and the archive of project documents contained in the document management system provide a detailed history of the changes that occur. **Change management** is the formal set of procedures, defined in the configuration management plan, that allows for additional changes to be considered and approved. Change management insures that all factors associated with a proposed change are evaluated, prevents unnecessary or marginal change, and establishes change priorities (Department of Defense Systems Management College, 2001).

Application of change management provides a number of benefits for a complex construction project (Project Management Institute, 2007):

- The correct version of all documentation is in use by the project team. In construction, a plan sheet may undergo many revisions. It is vital that all the project participants are using the same version.
- Change management insures that only authorized individuals make changes. It prevents unauthorized individuals from making ad hoc changes to the project baseline.
- Change management provides a planned method of notifying project participants of a change in the project configuration.
- A record of all changes is kept to support auditing and project closure activities.

Figure 10.2 shows a flowchart for how a change is managed using the configuration management technique. It shows how a change is formally processed and how it results in a modification of the project baseline.

Audits and Reviews

Complex projects require construction by different disciplines to be integrated together for the completed project to perform correctly. The auditing process is closely linked to

Figure 10.2

The change management process

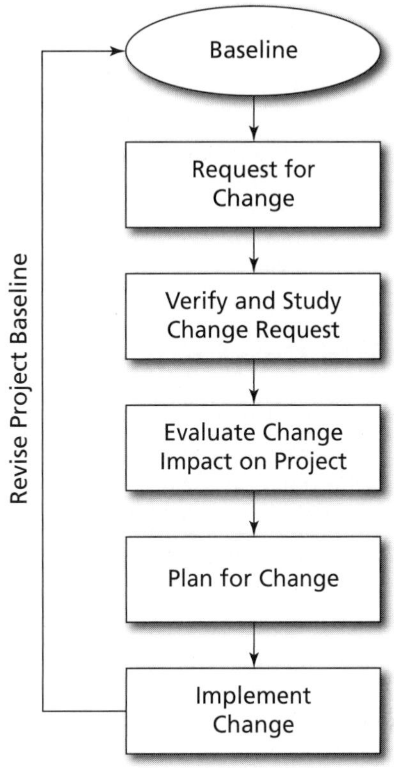

both baseline management and change management. The purpose of audits is to verify that installed systems are working properly and according to the specifications and validate the current state of a construction project.

There are different types of audits that can be conducted as part of a configuration management system (National Cooperative Highway Research Program, 2001). Two important types of audits are:

- **Functional configuration audits.** These audits are carried out to evaluate the testing of components of the construction project and the capability of installed components to perform their specified functions correctly. A configuration management system archives all the documentation related to the design and construction of project components and allows the owner to determine the current state of the project.
- **Physical configuration audits.** These audits evaluate the engineering drawings for accuracy. This is done to insure that the drawings reflect the latest project configuration. This type of audit insures that the completed as-built configuration conforms to the project baseline. On poorly managed projects, it has been found that it is sometimes difficult to identify what has actually been installed when the

project is complete. For a major project like a transit system, which will need to be operated and maintained for many decades after the completion of the project, any changes, if required, to the constructed facility would be difficult to implement if what was actually built is found not properly identified in the final documentation. Physical configuration audits prevent this from happening.

In essence, audits and reviews provide evidence to the system owner that everything is working according to specifications and has been installed according to specifications.

An Example Configuration Management System

Information technology can greatly facilitate the implementation of a configuration management system. One such system that has been developed specifically for use in the construction industry is a configuration control system by PACO Technologies called ccsNet. The computer system is Web based and provides the necessary capabilities to implement a configuration management system (PACO Technologies, 2007a). The system supports baseline management and change management, including the management of changes due to changes in the design and the construction contract. An agency that adopts the ccsNet system can configure it to reflect its business practices, and to conform to the agency's configuration management plan for governing inter-project communications, document storage, and change management (Steinberg and Otero, 2007).

The ccsNet system also provides configuration control of drawings and specifications. This becomes important during construction. There is a formal system in place to provide the latest design documents to contractors working in the field and to insure that everyone is using the same version of a document. The ccsNet system also provides the extensive document management capabilities required to implement a configuration management system. These capabilities include submittal tracking and review, requests for change, and action items.

The ccsNet system accepts computer documents such as Word, PDF, and CAD files. However, the system has the ability to accept documents that have been scanned. Typically, documents are scanned as PDF files. Another feature of the system is that it allows full text search within documents, including those that have been scanned, to make it easier for project managers to retrieve data. Each document entered into the ccsNet system receives a unique code number. An important feature is the ability to cross-reference documents to establish an audit trail of documents related to a particular change in the project.

Figure 10.3 shows the opening screen a user of the ccsNet would see after logging into the system. This screen can be used to display the user's tasks and correspondence. On this page, the user is able to see a listing of documents that require his or her review or approval. The documents have been automatically routed to the user in accordance

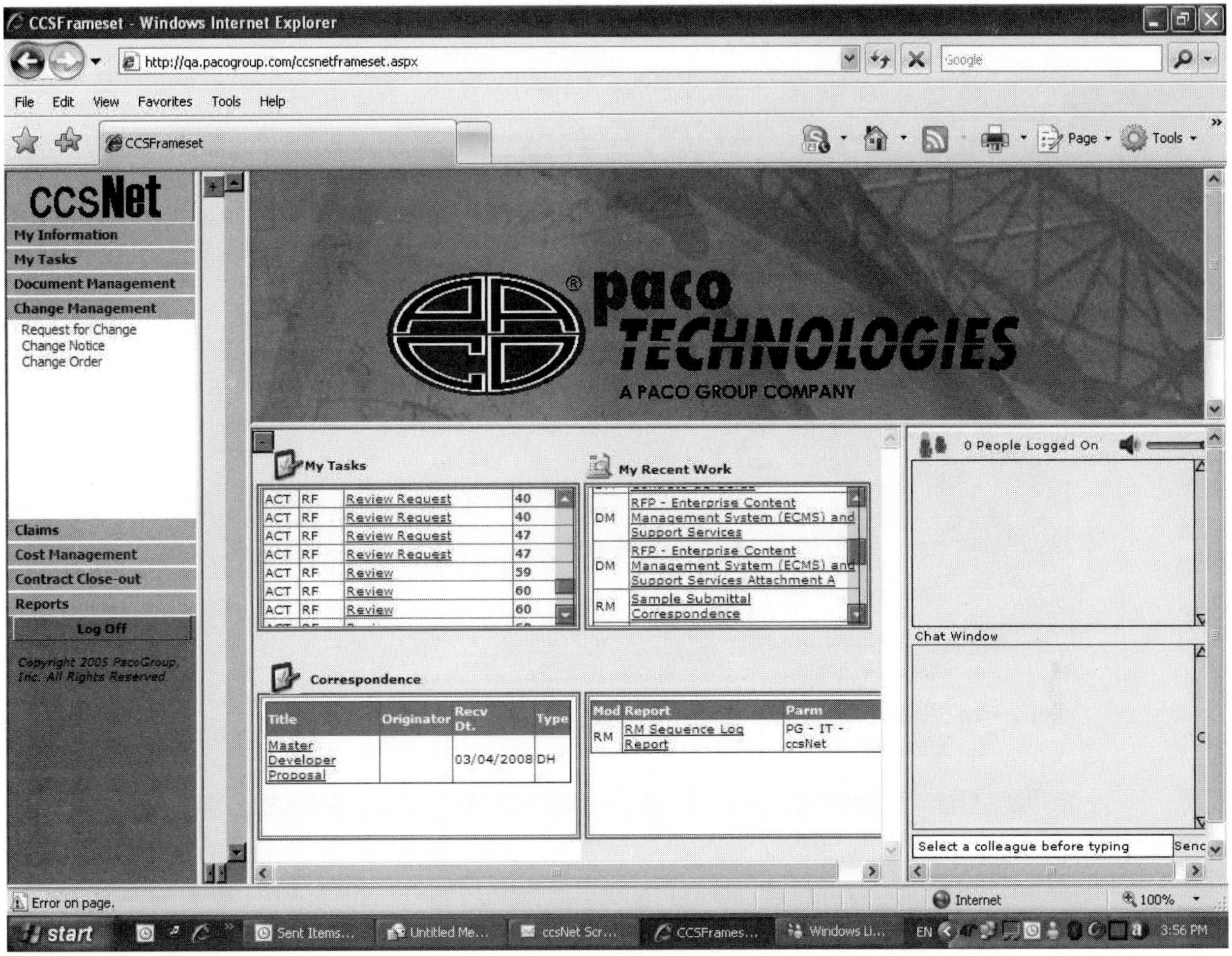

Figure 10.3 ccsNet opening screen. *Courtesy of PACO Technologies, a member company of PACO Group, Inc.*

with the configuration management plan. Figure 10.4 shows the form that is filled out to distribute a document to project participants through the document management system. The necessary information is collected on the form to route the submittal document to the person concerned on the project team for approval. Figure 10.5 shows a document management Web page that lists the various versions of a document. This allows a system user to go back and review the history of how an element of the project has had changes, as well as all related documentation.

Web-based configuration management software such as ccsNet provides a way of managing project documents and correspondence, and implementing the technique of configuration management to help control project changes, costs, and schedule.

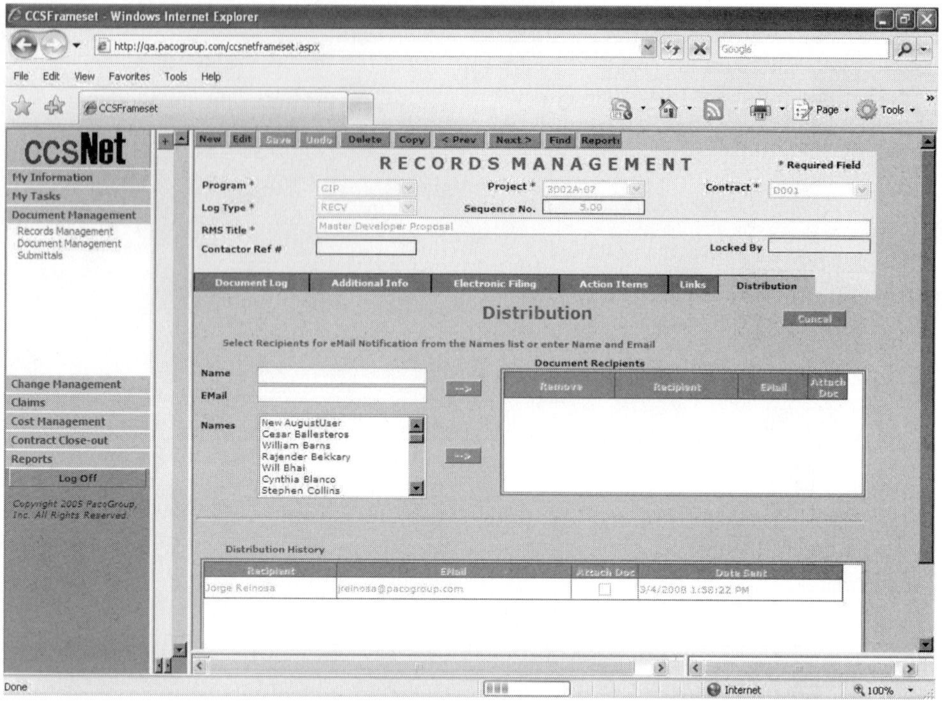

Figure 10.4 ccsNet form to distribute project records. *Courtesy of PACO Technologies, a member company of PACO Group, Inc.*

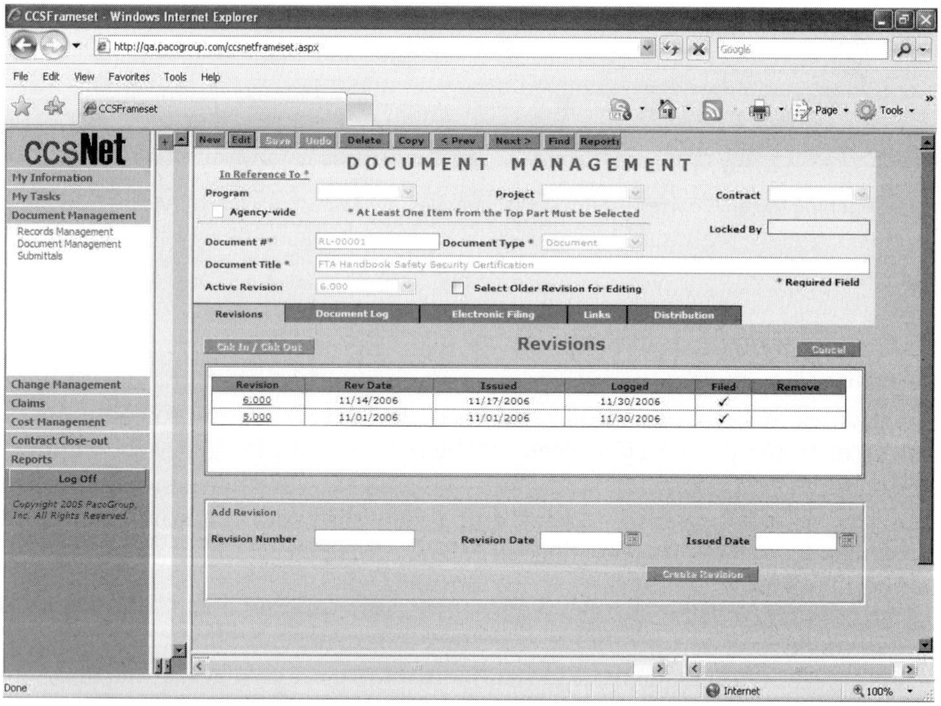

Figure 10.5 Using ccsNet to track documents. *Courtesy of PACO Technologies, a member company of PACO Group, Inc.*

Summary

Configuration management is emerging as a technique for managing complex construction projects. Construction projects change frequently from their conception through design and construction. There is a tendency for costs and schedules to change dramatically. Configuration management is a useful tool to manage this tendency toward cost increases and schedule slippages, because it provides a structured environment to evaluate and control project changes.

Key Terms

Baseline
Change management
Configuration management
Scope creep

Review Questions

1. Define configuration management.
2. What is a baseline? What causes a baseline to change?
3. What is change management? How does it prevent scope creep?
4. Why is document management so important to the implementation of configuration management?

Management Pro

Management Pro

Why did the Los Angeles Metro use configuration management on Red Line Segment 2? What were the benefits?

References

American National Standards Institute. 1998. *National Consensus Standard for Configuration Management*. ANSI/EDI 649-198. Washington, DC: American National Standards Institute.

Department of Defense Systems Management College. 2001. *Systems Engineering Fundamentals*. Fort Belvoir, VA: Defense Acquisition University Press. Available from http://www.dau.mil/pubs/pdf/SEFGuide%2001-01.pdf (accessed June 15, 2009).

National Cooperative Highway Research Program. 2001. *Configuration Management in Transportation Management Systems*. NCHRP Synthesis 294. Washington, DC: National Academy Press.

PACO Technologies. 2007a. Fact sheet: Configuration control system (ccsNet). Miami, FL: PACO Technologies.

PACO Technologies. 2007b. Fact sheet: Configuration management plan. Miami, FL: PACO Technologies.

PACO Technologies. 2007c. Fact sheet: Los Angeles County MTA case study. Miami, FL: PACO Technologies.

Project Management Institute. 2007. *Practice Standards for Project Configuration Management.* Newtown Square, PA: Project Management Institute.

Steinberg, Michael and Frank Otero. 2007. Using configuration management to mitigate the impact of design and construction contract changes. Paper presented at the Annual Meeting of the Construction Management Association of America, Chicago, IL.

chapter 11

Construction Productivity

Chapter Outline

Introduction

Previous chapters of the book have described how construction projects are organized contractually and how they are obtained. This book has also addressed how projects are planned and scheduled. This chapter considers how construction is accomplished in the field and how **productivity** is used as a measure of an output of construction work tasks. Economists define productivity as the amount of output that is produced per unit of input. For construction tasks, productivity is often recorded as the output per unit time or the output per man-hour. For example, earthmoving productivity is measured in cubic yards of soil moved per hour. The productivity of a crew of contractors could be given in board-feet per man-hour. A man-hour is a measure representing one person performing 1 hour of work.

A construction project includes many individual tasks that, when aggregated, result in the completion of a construction facility. These are tasks like "pour footings" or "build masonry wall." Chapter 7 discussed how these individual tasks are arranged in a schedule and duration is allocated to each task to calculate a CPM schedule. Determination of how long a task will take and the best way to accomplish the task requires a consideration of several management issues:

1. What equipment and labor are required to perform the construction task?
2. How long will it take a particular combination of manpower and machinery to perform a task?
3. If different groupings of people and equipment can be used to perform a task, what is the most cost-effective mix?

Given the management decisions that must be made, it can be seen that the manner in which a contractor decides to carry out different construction work tasks affects the productivity of the work task. Also, the productivity is linked to how much a work task will cost to perform. To perform a work task faster may require more equipment and/or manpower, increasing the project estimate.

Importantly, productivity is linked to how efficient and well managed a construction project is. Poor management can result in a poorly organized project with low productivity. This has led many owners to be critical of the construction industry; they feel that projects have cost more than they should have because of bad management, which causes low productivity. Productivity is a major concern in the construction industry.

Construction Productivity and the U.S. Economy

There was considerable concern over the decrease in productivity in the U.S. construction industry in the 1960s and 1970s (Building Futures Council, 2006). Owners, who pay for construction projects, naturally require that construction productivity be as high as possible so that their projects are completed in the most economical way. There is, however, little agreement between economists and industry experts about the actual trends in construction productivity over the last several decades.

One study by Allmon et al. (2000) looked at the labor productivity changes, as measured from cost estimating handbooks, for various construction activities over the period 1970–1998. This study found that construction productivity generally increased in the 1980s and 1990s.

Innovation

Innovation is the implementation of new methods, tools, or equipment that increases construction productivity. It is generally considered that increasing productivity through innovation will provide construction contractors the capability to be more competitive and profitable, while providing owner organizations the best value for their construction dollars. The construction industry is often criticized for resisting innovation. However, emerging technologies, such as automation of tasks handled by construction equipment and building of information-modeling tools to provide constructors with improved computer-based methods of visualizing construction projects, hold great promise for increasing construction productivity.

General Productivity Concepts

To measure the effectiveness of how a construction task is being performed, its productivity is measured.

Numerically, the output of a construction task is given as:

$$\text{Productivity} = \text{Output}/\text{time unit}$$

For example, if an earthmoving operation moved 20,000 cubic yards of soil in an 8-hour day, then the hourly productivity would be:

$$\text{Earthmoving productivity} = 20{,}000 \text{ cu yd/day}/8 \text{ hr/day} = 2{,}500 \text{ cu yd/hr}$$

This productivity information can then be used to derive the duration of the work task. If the contractor knows the quantity to be placed, then the work task duration will be:

$$\text{Task duration} = \text{Quantity of work}/\text{productivity}$$

Finally, if the cost per unit time for the equipment or crew is known, the cost for the work task is:

$$\text{Task cost} = \text{Task duration} \times \text{Cost/unit time}$$

If the earthmoving operation described in the preceding text must move a total of 200,000 cubic yards of soil, and the cost for trucks and loaders is \$3,000 per day, the earthmoving task will take 10 days and will cost \$30,000.

A clear link between the cost of a construction activity and its productivity has been shown. If two different methods of performing a construction task have the same unit cost, but one has a higher productivity, the higher productivity method will cost less and take a shorter amount of time.

Factors That Affect Productivity

When considering productivity, some construction work tasks are labor intensive whereas others are equipment intensive. For labor-intensive tasks, the productivity is determined by the motivation, skill, and training of the workforce, as well as the working conditions. For equipment-intensive operations, productivity is primarily dependent on the power and capabilities of the equipment used. Typically, building construction tasks such as installing drywall or plumbing are labor intensive because the job is mostly performed by human labor. Earthmoving work that relies on the speed and size of heavy machinery is equipment intensive.

Factors That Affect Labor Productivity

Both the project environment and the nature of the work that is to be performed affect labor productivity. Figure 11.1 illustrates how the inputs provided to a construction process (labor work-hours) are transformed into an output (completed construction work). The productivity of the construction process is affected by the influence of

Figure 11.1

Factors affecting labor output

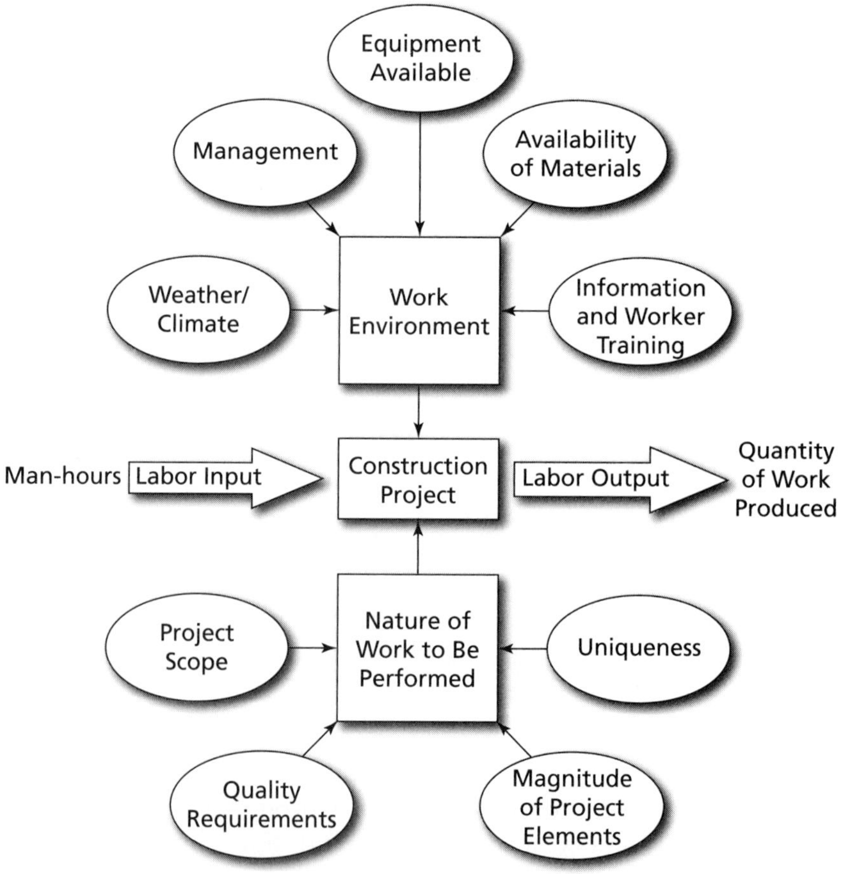

various factors. The important characteristics that affect labor productivity have been identified (Thomas and Sakaran, 1994). They focus on both what the construction site conditions are and the type of construction being performed. Factors that affect productivity include:

- **Work space and work flow.** Congestion at the construction site and improper sequencing of construction work tasks have a negative bearing on productivity.
- **Weather.** Poor weather conditions or a harsh climate adversely affect productivity.
- **Project management.** The quality of project management affects productivity: Poor management tends to decrease productivity.
- **Tooling and equipment.** The quality of the tools and equipment used affects productivity. Use of equipment that embodies the latest technological innovations tends to increase productivity.
- **Availability of raw materials.** The availability of construction materials is an important factor in productivity. Construction crews must be provided with a steady stream of materials to ensure high productivity.
- **Clear understanding of specifications.** Good information at the job site is vital to maximizing productivity. Workers must have a clear understanding of what is to be constructed and be well trained to achieve maximum productivity.

What is to be constructed will also affect productivity. Factors affecting productivity that are related to the nature of the work to be done include:

- **Uniqueness of the project.** The uniqueness of a project affects its productivity. An unusual building with many unusually shaped elements will naturally have lower productivity than a more common rectangular-shaped building.
- **Variation in scope.** Projects may vary greatly in scope. A megaproject may experience a different productivity level compared with a very small project.
- **Magnitude of the project.** The magnitude of what is being installed can affect productivity. For example, very large structural elements may be more difficult to install than smaller ones and will have lower productivity.
- **Quality.** The quality requirements of the construction will affect productivity. Specifications dictate quality requirements. Projects with specifications that require high quality will tend to have lower productivity than projects with less stringent quality requirements. In the former case, construction workers will need to be more exacting during the installation process, and this will take more time.

Measuring Labor Productivity in the Field

A simple way of measuring labor productivity is the 5-minute rating technique. It is based on the observation of a crew of workers over a short study period (Oglesby et al., 1989). The purpose of the 5-minute rating technique is to create awareness of where delays are occurring, what the magnitude of delays are, and if crews are operating in an effective manner.

A block of time is observed and if the worker is engaged in productive work for more then 50% of the time block, the worker is recorded as working. Using this technique, no block of time should be less than 5 minutes, with a rule of thumb that observations should include at least 1 minute for each crew member.

What will be recorded as work must be carefully defined for each study. Work includes effective work, which involves the installation of new equipment and materials in a project. Work also includes contributory work, that is, work that supports the project installation work. An example is workers building temporary scaffolding. An idle rating is given if the worker performs no effective or contributory work for more then 50% of the measured time block.

A 5-minute rating requires an observer with a watch and a form for recording observations. Table 11.1 shows a typical form for a 5-minute rating study. There are 11 time blocks studied. With three crew members, this means there is a total of 33 time blocks that can be recorded. During the study, work was recorded for 18 of the time blocks, giving an effectiveness of 55%. One problem observed during the study was that the laborer was not supplying the masons with bricks until the masons had run out of bricks. One way to improve the effectiveness of this crew would be for the laborer to bring more bricks before all the bricks at the workplace have been expended. The 5-minute survey has indicated a crew where management may want to intervene to introduce improved work procedures.

Table 11.1 A 5-minute rating of a masonry crew

TIME	MASON 1	MASON 2	LABORER 1	REMARKS
11:15	X	X		
11:16	X	X		
11:17	X		X	Masons ran out of bricks
11:18			X	
11:19			X	
11:20	X	X	X	
11:21	X	X		
11:22	X	X		
11:23				Masons ran out of bricks
11:24				
11:25	X	X	X	
Total = 33; working = 18; effectiveness = 55%				

X, effective work.

Human Learning

People learn as they perform repetitive activities. In construction, where many activities are repetitive, productivity increases as workers become more skilled in performing their tasks. Human learning has been extensively studied. When performing a task, it will take a shorter time the second time it is done. Each time the task is repeated, it takes a progressively shorter time but at a decreasing rate of improvement. Eventually, after several repetitions, there is a leveling off of productivity improvements.

Mathematical relationships have been developed that express the rate at which learning takes place. Learning curves typically take the shape shown in Figure 11.2. A model to predict the amount of time it will take to construct a unit of work is:

$$T_N = KN^S$$

where T_N is the man-hours or cost per unit of construction for the Nth unit, N the unit number, K the estimate for the first unit to be constructed, and S the slope of the improvement rate.

Using this model, it is assumed that the man-hours or cost necessary to complete a unit of construction will decrease by a constant percentage each time the quantity is doubled. In construction, a rate of improvement of 20% between doubled quantities is often assumed (Ostwald, 2001). This establishes an 80% rate of learning between doubled quantities. This means that the second unit will take 80% of the effort taken for the first unit and the fourth unit will take 80% of the effort of the second unit and 64% of the effort of the first unit.

Improvement is designated as ϕ. S, the slope of the improvement rate, is given by:

$$S = \frac{\log \phi}{\log 2}$$

Therefore, for 80% learning, S is:

$$S = \frac{\log 0.8}{\log 2} = -0.3219$$

Assume that three identical buildings are to be constructed. The first unit costs \$1,000,000. As there will be an 80% learning, the cost of the second unit will be:

$$T_N = \$1{,}000{,}000(2)^{-0.3219} = \$800{,}000$$

Figure 11.2

A learning curve

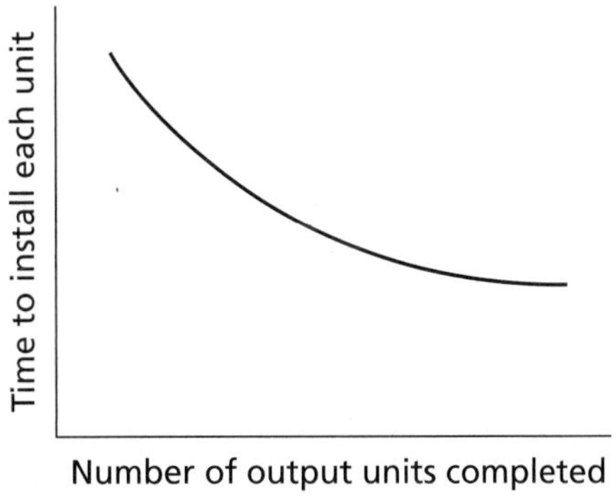

The cost of the third unit will be:

$$T_N = \$1{,}000{,}000(3)^{-0.3219} = \$702{,}100$$

Factors that Affect Equipment Productivity

Several factors dictate the level of productivity that construction equipment can provide. The levels of automation of a machine and its ability to amplify human energy have a large effect on the productivity that can be achieved.

Automation

Automation can be defined as the technique of making processes or machines self-acting or self-moving. Automation also pertains to the technique of making a device, machine, process, or procedure fully automatic. It appears that the area of construction that has so far benefited the most from automation is heavy construction. The type of work done in heavy construction, the placing and moving of bulk materials, can be more easily automated than some of the labor-intensive tasks found in building construction that will require advanced robots to implement.

The implementation of automation can have impacts beyond technical issues. Automation will increase productivity, but it can also displace workers from their jobs because fewer or less-skilled workers are required after automation is implemented. Consideration should be given to the effect on the labor force before implementing a program of automation.

There are several levels at which automation can be implemented. At the first level, equipment can be added to construction machinery to provide improved feedback to the equipment operator. At the second level, computerized equipment can be incorporated into the construction equipment to enable continued operation with little or no operator intervention. Finally, robotic equipment can be employed that is completely autonomous. There are many motivating factors that drive the widespread adoption of construction automation. These include (Khoshnevis, 2004):

- **Competition.** Automation offers the possibility of gaining an advantage over competitors using lesser automation.
- **Labor inefficiency.** In some types of construction labor, efficiency is low.
- **Shortages of skilled labor.** Automation can reduce the need for labor in markets where there are labor shortages.
- **Safety.** Automated equipment can be employed in hazardous environments, thereby not exposing people to unsafe situations.
- **Practical difficulties.** It is difficult to manage and control at the construction site.
- **Technological advances.** Like many technologies discussed in this book, new advances make automation a possibility for different types of construction operations that would have been impossible to contemplate even a few years ago.
- **Consistent quality.** Automation improves not only construction work quality but also minimizes or eliminates product variation.

As seen from preceding text, there are many areas where automation can be beneficial, with the potential to reduce operation cycle times and increase productivity.

Robots

There are various definitions of what a **robot** is. However, it has been suggested that a robot must have several essential characteristics (MacDonald):

- The robot must possess some form of mobility.
- It should be programmable to perform a variety of tasks.
- It should operate automatically after programming.

Researchers in construction have experimented with the use of robots in the construction industry, and there have been some commercial applications of robots. However, the high cost of robots and their complexity have been a barrier to their implementation in construction. Construction is fundamentally different from manufacturing, where a robot can be installed on an assembly line and then left to work for years without requiring movement. However, in construction, projects can be relatively short lived, and the robot must be moved to the project and also moved within the project (such as between floors on a high-rise building).

Warszawski and Navon (1998) noted that conventional work in building construction is adapted to manual work, where a construction worker uses various materials and tools to complete a task. The complex motions of a person (finding, picking, placing, and attaching) and use of multiple tools are difficult for a robot to perform. They suggest that consideration must be given during design to use simplified building materials and systems that a robot can construct. Care must also be given to design a structure where robots can move easily about the exterior and interior of a structure.

In the last 15 years, several prototype robots have been developed. These earlier robotic implementations can be grouped into several areas, two of which are:

- **Materials handling.** Robots employed in lifting large loads.
- **Finishing robots.** Robots employed in tasks such as spray painting and concrete finishing, both in the exterior and interior of a building.

Japanese construction companies have been active in implementing robotic technologies. Japanese companies employed robots because of labor shortages, the desire to remain competitive internationally, and the desire to market their use of advanced technology to potential clients. Additionally, large Japanese contractors have shown more willingness to invest funds for research and development than their American counterparts (Everett and Saito, 1996). An example of the automation and robotic building construction techniques that have been employed in Japan is the SMART system (Kangari and Miyatake, 1997). The system was employed to automate the construction of high-rise buildings. This technology included automated transportation systems, automated welding, placing of floor slabs, and an integrated information management system. Extensive use was made of prefabricated components and simplified joints to facilitate the use of robots and automation.

With the practical application of robotic techniques to construction in Japan in the 1990s, there was considerable interest in robotics shown by the U.S. construction research community. Although many prototype robotic systems have been developed, few commercial applications have evolved. The economic downturn that occurred in Japan in the late 1990s also served to "cool" the environment for research into robotics in building construction. However, research continues, and someday, with decreased costs and increased computing power, robots will become more commonplace. Therefore, it is suggested that construction managers should monitor developments in this area.

Some of the recent work in robotics is summarized in the following list:

- Work at the National Institute of Standards and Technology (NIST) has focused on the development of a test bed for researching robotic structural steel placement (Lyttle et al., 2004). NIST has experimented with a robotic crane fitted with a laser-based 3D site measurement system. The system is capable of autonomous path planning and navigation. Figure 11.3 shows an autonomous steel docking using a robotic crane. More details about this research are available on NIST's Building and Fire Research Laboratory Web page (http://www.bfrl.nist.gov/).
- Recent research has been performed concerning the possibilities of transferring robotic techniques from manufacturing to the construction industry. Khoshnevis (2004) has discussed the potential to use contour crafting, an automated layered fabrication technology, to construct whole houses and their subcomponents in a single run.

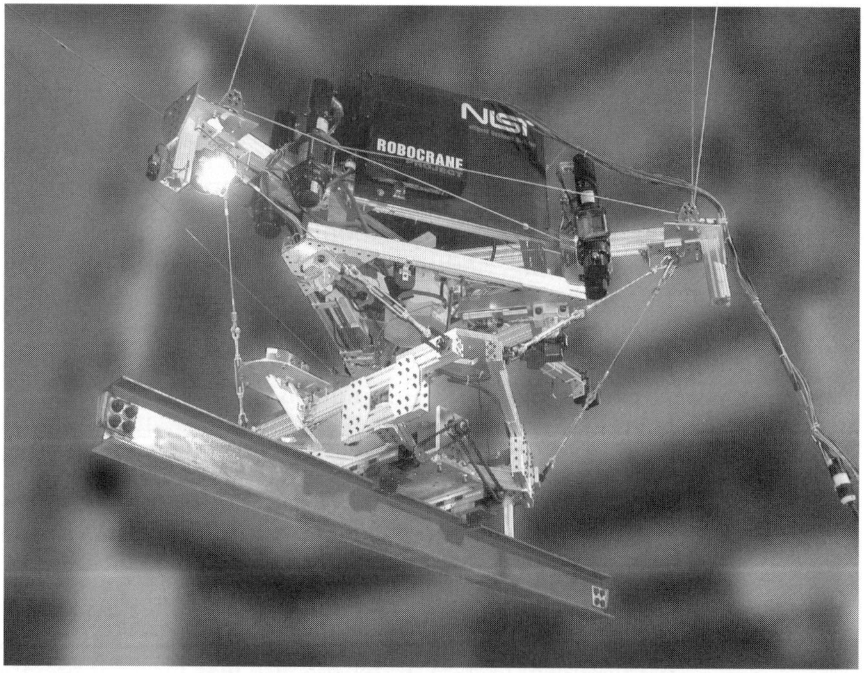

Figure 11.3

RoboCrane for steel erection. *Courtesy of NIST.*

Some Practical Considerations for Determining Equipment Productivity

Many factors at the construction site affect the productivity of construction equipment. An important use of construction equipment is earthmoving and the manipulation of large volumes of soil. In heavy construction, this may involve moving huge quantities of rock for the construction of a dam or a road. For building construction, the building site needs to be graded and building foundations need to be dug.

Soil and Rock Conditions

Soil and rock conditions that are encountered at a construction site can vary significantly between projects in different geographical locations. The U.S. National Resources Conservation Service (2008) has identified more than 20,000 unique soils in the United States. The properties of soils can vary widely. The type of soil found at a particular project site will influence construction techniques and equipment that can be used. In addition, the moisture content of soil can significantly affect its properties. Soils like loose sands or clay with high moisture content, although differing significantly in physiochemical properties, can be difficult to work in. Obviously, if construction equipment becomes "bogged down," productivity will decline.

Different techniques can be used to improve soil conditions for a project. Physical compaction of the soil is often possible. There are construction compaction machines that can be used to increase the density of the soil. The type of machine that is most effective varies by the nature of the soil. For example, a smooth-wheeled roller works well in compacting clean gravel, whereas a tamping foot roller works well in clay. When conditions of haul roads within a construction site are particularly poor, contractors are often able to add chemicals to the soil to strengthen it and allow construction equipment to traverse the site more rapidly. The process of mixing additives with soil to improve its engineering properties is called soil stabilization. Chemicals used are typically lime, cement, asphalt, and fly ash.

Equipment Power and Selection

The power of the construction equipment selected determines to a large extent the speed and productivity that will be achieved with the machine. The power required from a machine is a function of its internal friction, friction between the wheels (or tracks) and the roadway, and the grade over which the equipment will operate. Two factors establish the power required for a construction task.

Rolling resistance is the external force that resists the motion of wheeled vehicles. The engine must supply the power to overcome this resistance. The rolling resistance for wheeled construction vehicles is determined by the weight of the vehicle, the roadway characteristics, and the amount the vehicle tires penetrate the roadway. Different soil and

Table 11.2 Rolling resistance factors

ROADWAY TYPE	ROLLING RESISTANCE, lb/ton	ROLLING RESISTANCE, lb/lb
Hard-smooth stabilized, surfaced roadway	40	0.020
Earth roadway, rutted, flexing under load	100	0.5
Loose sand or gravel	200	1.00
Soft muddy, rutted roadway	300–400	1.50–2.00

roadway types will have widely varying rolling resistances. Table 11.2 shows the rolling resistance for several different conditions (Sain and Quinby, 2004).

The formula for the **rolling resistance,** R, of construction equipment with tires is:

$$R = (R_{\text{factor}} + R_{\text{penetration}}P)\ W$$

where R_{factor} is the rolling resistance factor. Assume a value of 40 lb/ton for this factor and 30 lb/ton for $R_{\text{penetration}}$, the tire penetration factor. P is the number of inches of tire penetration on the road, and W is the weight on the wheels of the vehicle. Therefore, the rolling resistance of a 50-ton truck with tires penetrating a muddy roadway by 2 inches would be:

$$R = (40 + 2(30))\ \text{lb/ton} \times 50\ \text{tons} = 5{,}000\ \text{lb}$$

The other component of the total resistance, the **grade resistance,** increases on upgrades and decreases on downgrades. The grade resistance is given by:

$$G = R_{\text{gf}}\, gW$$

where G is the grade resistance in pounds, g the grade in percentage, and W the weight on wheels in tons. R_{gf} is the grade resistance factor and has been found to be 20 lb/ton/%grade. Therefore, the grade resistance of a truck with a 50-ton weight on its wheels on a grade of 3% is:

$$G = 20\ \text{lb/ton/\%grade} \times 3\% \times 50\ \text{tons} = 3{,}000\ \text{lb}$$

The **total resistance,** TR, is:

$$\text{TR} = R + G$$

Equipment must be found that can produce power that is greater than the total resistance. Equipment manufacturers provide tables that are helpful in assessing the power output of their equipment and selecting the correct equipment model for the task at hand. A typical chart is shown in Figure 11.4. This chart is for a Caterpillar 730 articulated truck. The articulated truck is shown in Figure 11.5. The chart works by converting the rolling resistance in pounds to an **equivalent grade:**

$$\text{Equivalent grade} = \frac{R}{20\ \text{lb/ton/\%grade}}$$

Figure 11.4

Performance curves for articulated truck. *Reprinted courtesy of Caterpillar Inc.*

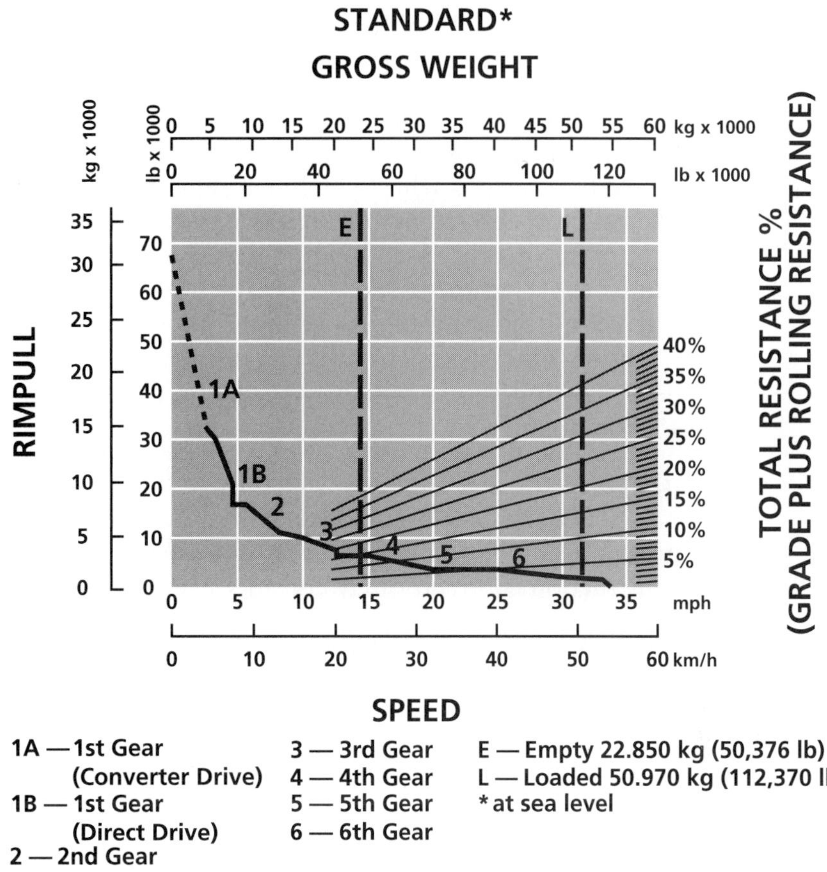

Figure 11.5

Caterpillar 730 articulated truck. *Reprinted courtesy of Caterpillar Inc.*

Then, the total resistance is:

$$TR = \text{Equivalent grade} + \text{Actual grade}$$

The following example illustrates how to use the chart. Assume that the vehicle is loaded and the total resistance is 10%. Read down from the loaded weight at the top of the chart to where the line intersects the sloping line representing 10% total resistance. Then move horizontally to the numbered gear with the highest attainable speed. In this case, it is third gear. Reading down from the third gear curve, it can be seen that the truck will be able to operate at a speed of about 9.5 miles/hour. The **rimpull** generated can be determined by moving horizontally to the left axis from the third gear. Rimpull is defined as the maximum amount of pull that can be generated at the wheel rims of a wheeled vehicle. It will be approximately 11,000 pounds.

If, upon reading this curve, the construction equipment produces insufficient speed, it is necessary to seek a model with higher power output. In addition, if a machine produces enough power but only in a very low gear, where it will operate slowly, a construction contractor may consider using a more powerful equipment.

The speed determined using the chart is the maximum speed at which the vehicle will operate. To determine travel times, acceleration and deceleration of the machine must also be considered. Equipment manufacturers often provide equipment purchasers with tables that can provide total travel times, including deceleration and acceleration for varying haul lengths.

Operating conditions at the construction site often make it impossible to use all the power that a construction equipment is able to generate. In particular, in slippery or frozen conditions, where traction is poor, a vehicle cannot use all the power available to it as a driving force. The factors that determine how much force can be developed in adverse surface conditions are the coefficient of traction of the ground or road surface and the weight on the driving wheels of the vehicle. For tracked vehicles, the total weight is used in the calculation. Table 11.3 shows some typical coefficients of traction for wheeled and tracked vehicles (Sain and Quinby, 2004).

Table 11.3 illustrates why tracked vehicles can offer superior performance in muddy conditions. Tracked vehicles can generate higher traction coefficients in poor soil conditions. The power that a machine can deliver is calculated as:

$$\text{Deliverable power} = \text{Coefficient of traction} \times \text{Weight on driving wheels}$$

The weight on the front wheels of the articulated truck discussed in preceding text is 34,965 pounds. Therefore, the deliverable power that can be generated is:

$$\text{Deliverable power} = 0.45 \times 34{,}965 \text{ lb} = 15{,}734.25$$

Note that the deliverable power is greater than the required power calculated using the performance chart.

Table 11.3 Coefficient of traction for various conditions

SURFACE TYPE	COEFFICIENT OF TRACTION FOR RUBBER-TIRED VEHICLE	COEFFICIENT OF TRACTION FOR TRACKED VEHICLE
Concrete	0.90	0.45
Rutted clay loam	0.40	0.70
Gravel road	0.36	0.50
Ice	0.12	0.12
Loose earth	0.45	0.60
Firm earth	0.55	0.90

Figure 11.6 summarizes the process of selecting construction machinery with enough power. First, the power requirement dictated by the rolling resistance and grade at the site must be determined. Second, the operating environment defines the useable power that can be used. Finally, site roadway improvements must be considered to increase the useable power.

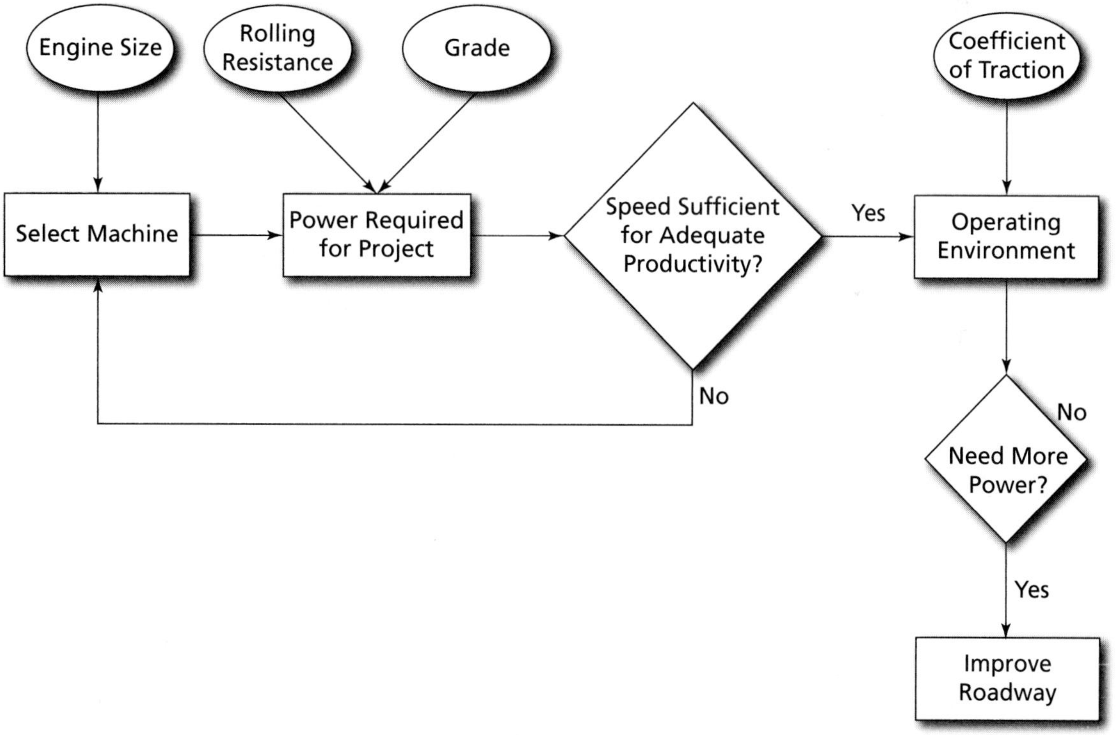

Figure 11-6 Selection of equipment

Matching the Productivity of Different Equipment Types: The Balance Point

Many construction tasks require the interaction of two different types of equipment. The equipment will have differing productivities. The question arises of how many of one type of equipment (with a lower productivity) should be matched with the other equipment type (which has a higher productivity). A common example is how many trucks to use with a front end loader. The productivity of the single loader must be matched with the total productivity of several trucks to achieve a balance between the productivities of the two equipment types. The productivity where this is achieved is called the **balance point**.

Balance Point Example

Figure 11.7 shows a production system where a loader is loading trucks. The loader fills each truck with 7 cubic yards of soil. The trucks travel to a dump site, dump, and return to be loaded again. The loader is either filling a truck with soil or getting more soil from a spoil bank. Table 11.4 shows the times required for each operation observed in the field. Notice that the trucks take less time to return because they are empty. To determine the productivity of a truck, the number of cycles that can be completed in 1 hour must be determined. This is:

$$60 \text{ min/hr}/19 \text{ min/cycle} = 3.16 \text{ loads/hr}$$

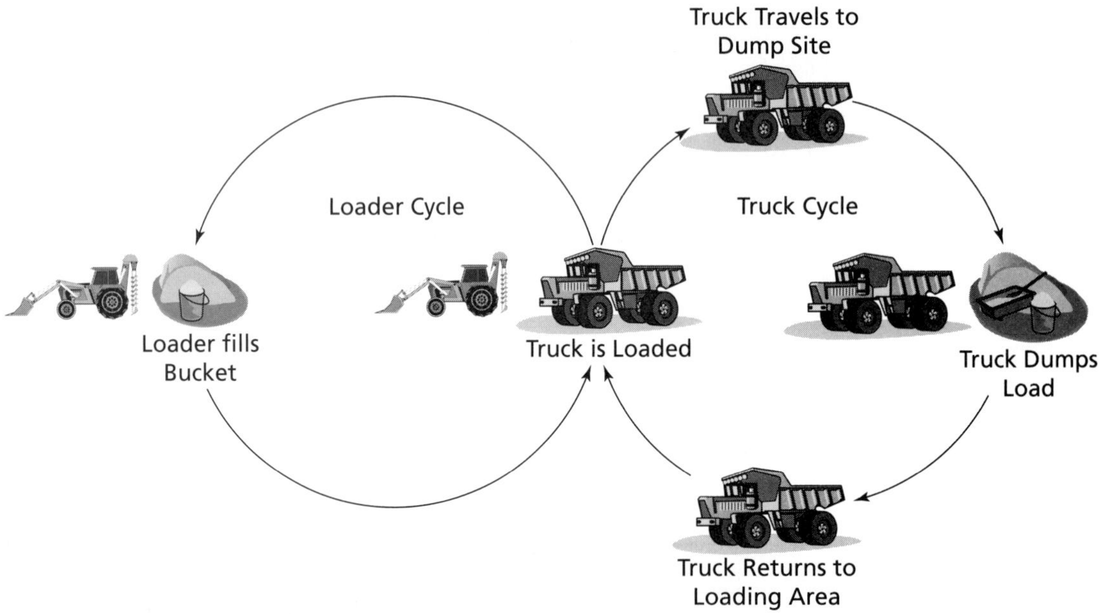

Figure 11-7 Balance point analysis

Table 11.4 Activity durations

ACTIVITY	DURATION (MINUTES)
Loader fills bucket	1.5
Load trucks	1.5
Travel to dump site	9
Dump soil	1.0
Trucks return to be loaded	7.5

The productivity of a single truck will be the number of loads per hour multiplied by the amount of soil hauled per load:

$$3.16 \text{ loads/hr} \times 7 \text{ cu yd/load} = 22.11 \text{ cu yd/hr}$$

The productivity of the loader will be:

$$60 \text{ min/hr}/3 \text{ min/cycle} \times 7 \text{ cu yd/load} = 140 \text{ cu yd/hr}$$

The balance point is the loader productivity divided by the truck productivity:

$$140/22.11 = 6.33$$

The balance point gives the number of trucks that will balance the productivity of the loader. In an ideal system, this would represent the point where all the equipment is productive all the time and there is no wasted productivity. However, construction equipment exist only as discreet units: Only six or seven trucks can be utilized. In this example, six trucks would be selected to work with the loader.

Balance Point Assumptions

The balance point is to some extent an idealized analysis. The balance point analysis is based on the idea that each truck-and-loader cycle is of a constant duration. It is widely known in the construction industry that time will often vary because of equipment breakdowns, operating conditions, or other unforeseen factors. Therefore, it should be expected that real productivity achieved in the field will be less. To achieve more realistic estimates of construction productivity, computer simulation can be employed. Using computer simulation, the various elements of the loader and truck cycle times can be represented as statistical distributions rather than fixed durations. This allows for the variability inherent in construction activities to be included in the productivity analysis.

Summary

In this chapter, construction productivity has been discussed. There has been considerable discussion in the construction industry that productivity may have decreased in the 1960s and 1970s. Innovation is seen as one way to increase construction productivity by employing information technology and automation in the industry.

In the context of labor productivity, this chapter has discussed various factors such as the conditions of the project site and the nature of what is to be built that can significantly affect productivity. Human learning has shown that time reductions occur as construction crews repeat the task. Regarding equipment, it has been shown that equipment productivity varies because of the power generated by the machine selected for use, the machines capacity and conditions at the construction site. For some repetitive equipment-intensive construction tasks, a balance point can be found that maximizes equipment usage.

Key Terms

Automation	Grade resistance	Robot
Balance point	Productivity	Rolling resistance
Equivalent grade	Rimpull	Total resistance

Review Questions

1. A truck with a 5-cubic-yard capacity is filled by a loader. This process takes 3 minutes. The truck travels to the site of a highway project where it dumps its load as fill for the roadway base course. It takes 2 minutes to dump the load. The truck returns to the loading area (duration = 5 minutes). The total cycle time for the loader is 4.5 minutes. How many trucks should be employed to maximize equipment usage?
2. A crew of five masons takes 8 hours to build a wall consisting of 1,000 blocks. Express the productivity in terms of blocks per man-hour.
3. A backhoe digs a trench that is 100 feet long in 6 hours. How many feet of trench can the backhoe dig in a 40-hour week?
4. Six similar tunnels must be constructed on a major highway project. If the first tunnel costs $3,000,000, find the cost for the second and fourth tunnels if the learning rate is 80%.
5. An empty Caterpillar 730 articulated truck must traverse an equivalent grade of 15%. How fast can it go? What gear is used?

Management Pro

Management Pro

Find a construction project in your area and identify a working crew. Perform a 5-minute rating of the crew. Note why and when delays occurred. What is the effectiveness of the crew? Does the 5-minute rating indicate that the crew should be studied more? Write a short report describing modifications to the process that could reduce any delays observed.

References

Allmon, Eric, Carl T. Haas, John D. Borcherding, and Paul M. Goodrum. 2000. U.S. construction labor productivity trends, 1970–1998. *Journal of Construction Engineering and Management* 126 (2): 97–104.

Building Futures Council. 2006. Measuring innovation and evaluating innovation in the U.S. construction industry. Arlington, VA: Building Futures Council. Available from http://www.thebfc.com/pdf/BFC_Productivity_whitepaper.pdf (accessed March 15, 2008).

Everett, John G. and Hiroshi Saito. 1996. Construction automation: Demands and satisfiers in the United States and Japan. *Journal of Construction Engineering and Management* 122: 147–151.

Kangari, Roozbeh and Yasuyoshi Miyatake. 1997. Developing and managing innovative construction technologies in Japan. *Journal of Construction Engineering and Management* 123 (1): 72–78.

Khoshnevis, Behrokh. 2004. Automated construction by contour crafting-related robotics and information technologies. *Automation in Construction* 13: 5–19.

Lyttle, Alan M., Kamal S. Saidi, Roger V. Bostleman, William C. Stone, and Nicholas A. Scott. 2004. Adapting a teleoperated device for autonomous control using three-dimensional positioning sensors: Experiences with the NIST RoboCrane. *Automation in Construction* 13: 101–118.

MacDonald, Chris. What is a robot? Available from http://www.ethicsweb.ca/robots/whatisarobot.htm (accessed December 28, 2005).

Oglesby, Clarkson, Henry Parker, and Gregory Howell. 1989. *Productivity Improvement in Construction.* New York: McGraw-Hill.

Sain, Charles H. and G. William Quinby. 2004. In *Standard Handbook for Civil Engineers*, 5th ed. J.T. Ricketts, M.K. Loftin, and F.S. Merritt, editors. New York: McGraw-Hill. pp. 13.13–13.18.

Thomas, H.R. and Ahmet S. Sakaran. 1994. Forecasting labor productivity using factor model. *Journal of Construction Engineering and Management* 120 (1): 228–239.

U.S. National Resources Conservation Service. 2008. Soil formation and classification. Available from http://soils.usda.gov/education/facts/formation.html (accessed May 12, 2008).

Warszawski, A. and R. Navon. 1998. Implementation of robotics in building: Current status and future prospects. *Journal of Construction Engineering and Management* 124 (1): 31–41.

chapter 12

Construction Safety

Chapter Outline

Introduction

Construction can be a dangerous enterprise. Every year, there are many deaths and injuries on construction projects. Table 12.1 shows the deaths that occurred in construction in 2007 in the United States (Bureau of Labor Statistics, 2008a). Construction is the industry group with the highest number of fatalities with 1113 deaths. In comparison, manufacturing had 392 workplace fatalities in 2007. Construction had about 21% of all workplace accidents in 2007. Clearly, safety is an important issue in the construction industry.

Figure 12.1 shows fatalities by type. Note that falls are the largest component of construction fatalities, with transportation incidents the second largest. Falls occur frequently because construction workers often work on roofs and in unfinished buildings that may have many openings where a fall is possible. Transportation accidents are high because highway work zones for highway construction have proven to be a dangerous environment. Workers are not only struck by equipment on the project site but also exposed to the danger of motorists passing close by the work zone.

Construction activity also has many nonfatal accidents. In 2006, there were 412,900 injuries and illnesses in the U.S construction industry. Of this total, 153,200 were injuries that required days away from work (Bureau of Labor Statistics, 2008b).

The purpose of this chapter is to show not only that contractors have a moral obligation to provide a safe workplace but also that safe operations significantly reduce a contractor's costs. Accidents significantly increase project costs, both directly through increases in insurance premiums and indirectly because of the many non-reimbursed and hidden costs of accidents.

Table 12.1 Construction fatalities in the United States[a]

CONSTRUCTION TYPE	TOTAL FATALITIES
Construction of buildings	236
Heavy and civil engineering construction	216
Specialty trade contractors	238
Building equipment contractors	169
Building finishing contractors	122
Other specialty trade contractors	132
Total	1,178

[a] Figures are for the year 2007 (Bureau of Labor Statistics, 2008a).

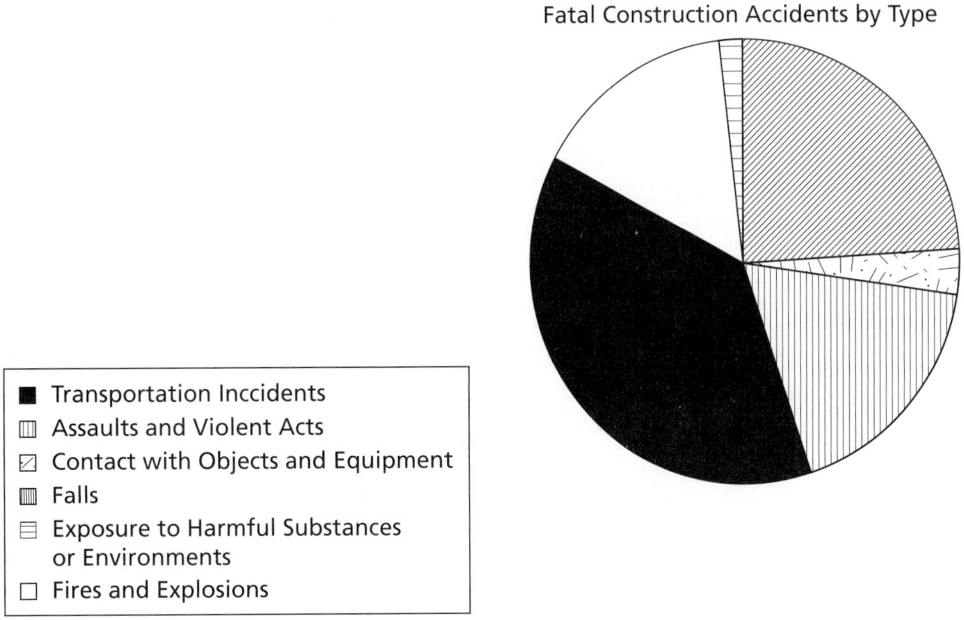

Figure 12.1

Pie chart illustrating fatality causes

Insurance Costs and Safety

One of the major effects of construction accidents is that contractors with high accident rates pay significantly more for insurance. This can significantly affect the contractor's competitiveness. Insurance rates are based on a contractor's safety record. **Workers compensation insurance**, which is paid by the contractor, is determined by the contractor's safety record. This type of insurance compensates workers that have been injured in the workplace. The rates for these insurances are quoted in dollars of premium per $100 of payroll. Each state has different rates, and the rates vary depending on the risk involved in the work.

Every contractor has an **experience modification rating** (EMR). It is a statistical tool used by the insurance industry to determine insurance rates. It has two purposes. First, the rating is used to tailor the contractor's insurance premium to the insurance carrier's cost of providing insurance. Employers who have more claims will pay more, whereas those with few claims will pay less. The second purpose is to provide construction contractors with incentives to operate in a safe manner and to provide a safer workplace to their employees. Insurance companies mandate that all contractors participate in the rating system. In most states, experience modifications are calculated by the National Council on Compensation Insurance (NCCI). The NCCI is a private, not-for-profit corporation, originally established and funded by the insurance industry. It provides the actuarial and rule-making framework that is the underpinning of workers compensation pricing (Safety Management Group, 2002).

An EMR of 1.00 represents the average claims rate for an industry. A contractor with a good safety record might have a rating of 0.65, which means the contractor has 35% claims less than the industry average. As part of a contractor's overhead, the company's cost for workers compensation insurance is passed along to the customer. The less the contractor spends on workers compensation, the less the customer will have to pay. A lower experience modification rate can result in huge savings. Contractors with poor safety records can pay more than double for workers compensation insurance.

Example

One example shows that ". . . a contractor with a $15-an-hour labor rate and an EMR of 0.60 can save $1.2 million in premiums on a $10-million job compared with a contractor with an EMR of 1.4" (Engineering News Record, 1991). The EMR allows construction contractors with a better safety record to provide a more competitive bid than contractors with a high EMR rating.

Levitt and Samuelson (1993) have calculated that general contractors with an EMR of 1.50 have a "handicap" in bidding against competitors with a better safety record, who have a 0.50 EMR totaling 2.2% of total revenue. The EMR rating allows safe contractors with the ability to control their insurance costs. A good safety record indicates a company is well managed and dedicated to keeping its operating costs low.

Uninsured Costs of Accidents

The effects of accidents go beyond increases in insurance premiums. There are many nonreimbursed costs and also many intangible costs associated with a poor safety record. These costs include man-hours spent in clearing an accident area or the cost of replacing damaged equipment. Productivity losses occur because of disruption to the project. They include construction output spoiled by the accident, the loss of skill and experience on the project, and lower output by a less-skilled replacement. Intangibles that can slow a project are decreased worker morale, increased labor conflict, and unfavorable public perceptions of the accident (Halpin and Woodhead, 1998). Additional costs may be the loss of a bonus for early completion and the overhead costs the contractor incurs while work is stopped. Clearly, the occurrence of an accident is costly for the contractor because of the disruptions and delays that will result on the project after the accident.

Self-Insurance

Some large construction companies have the capability to **self-insure**. Rather than taking a policy from an insurance company, the firm pays premiums into an escrow account and administers the claims itself. The benefit of self-insurance is that firms become safety conscious because they directly benefit from savings due to safety performance. It can be observed in self-insured companies that they have extensive safety programs and safety training (Levitt and Samuelson, 1993). It was observed on one self-insured project that a safety engineer was on site for the contractor, and that the safety engineer made frequent patrols of the site to find and report unsafe conditions.

Organizational Accidents and Construction

Reason (1997) has described a theory of accidents where "organizational accidents entail the breaching of the barriers and safeguards that separate damaging and injurious hazards from vulnerable people or assets." This is often called the "Swiss-cheese" theory of accidents where hazards break through holes in the defensive barriers to cause an accident. Therefore, accidents are reduced if the holes in the barriers can be reduced. In the construction industry, the defenses (barriers) to an accident include "hard" defenses such as personal protective equipment and physical barriers as well as "soft defenses" such as regulatory surveillance, training, and supervisory oversight. To avoid breaching of the barriers and to reduce accidents, this theory indicates that it is government regulation, combined with safety training of project workers and a management that encourages safe practices, that is necessary to reduce construction accidents.

Government Safety Regulations

Unsafe operation by a contractor can have significant regulatory impact in the form of fines for violation of federal safety rules. The **Occupational Safety and Health Administration** (**OSHA**) is the federal agency responsible for workplace safety in the United States. The Occupational Safety and Health Act, which created OSHA, was passed in 1970. OSHA is a part of the U.S. Department of Labor.

OSHA has several functions, some of which are presented in the following list (Princeton University, 2007):

- Encourage employers and employees to reduce workplace hazards and to implement new or improve existing safety and health standards
- Establish separate but dependent responsibilities and rights for employers and employees for the achievement of better safety and health conditions
- Maintain a reporting and record-keeping system to monitor job-related injuries and illnesses
- Establish training programs to increase the number and competence of occupational safety and health personnel
- Develop mandatory job safety and health standards and enforce them effectively

OSHA Enforcement

Every construction project is subject to inspection by OSHA. OSHA compliance offices are responsible for determining if a project is in compliance with OSHA regulations.

An inspection is initiated by OSHA for several reasons, some of which are presented in the following list (Hinze, 1997):

- Knowledge of imminent danger that could cause death or bodily harm
- Random selection of a work site for inspection

- A complaint made by a worker of an alleged violation of OSHA standards
- The occurrence of a fatality or an incident that caused injuries requiring hospitalization to three or more workers

An OSHA inspection begins with an opening conference between the contractor's representatives and the compliance officer. An employee of the contractor is selected to accompany the compliance officer on the site visit. Unions may also designate a representative to accompany the compliance officer. First, the injury records are checked to make sure they are current. This is the OSHA form 200 where accidents are reported. As the inspection proceeds, the compliance officer observes violations and shows them to the contractor's representative. The compliance officer makes sketches, takes photographs, and takes notes to document and understand the conditions on the site (Hinze, 1997).

At the conclusion of an OSHA inspection, the compliance officer meets with the contractor's representative in an informal meeting. All violations that were found are discussed in the meeting and contractor personnel are encouraged to ask questions about the violations. The compliance officer typically specifies the abatement period for each violation. Very severe safety problems can cause the compliance officer to shut down portions of the work until the safety deficiencies have been corrected.

After the compliance officer reports findings, the OSHA area director determines what citations, if any, will be issued and what penalties, if any, will be proposed. Citations inform the employer and employees of the regulations and standards alleged to have been violated and of the proposed length of time set for their abatement. The employer will receive citations and notices of proposed penalties by certified mail. The employer must post a copy of each citation at or near the place a violation occurred for three days or until the violation is abated, whichever is longer (U.S. Department of Labor, 1996). Table 12.2 describes the fines that can be imposed.

Table 12.2 OSHA violations and penalties

TYPE OF VIOLATION	PENALTY FOR EACH VIOLATION (in $)
Other than serious	Up to 7,000
Serious	Mandatory, up to 7,000
Willful	5,000–70,000
Repeated	Up to 70,000
Failure to abate prior violation	Up to 7,000 per day

The following are the types of violations, summarized in Table 12.2, that may be cited and the penalties that may be proposed (U.S. Department of Labor, 1996):

- **Other-than-serious violation.** A violation that has a direct relationship to job safety and health, but probably would not cause death or serious physical harm. A proposed penalty of up to $7,000 for each violation is discretionary. A penalty for an other-than-serious violation may be adjusted downward by as much as 95%, depending on the employer's good faith (demonstrated efforts to comply with the Act), history of previous violations, and size of business. When the adjusted penalty amounts to less than $100, no penalty is proposed.
- **Serious violation.** A violation where there is substantial probability that death or serious physical harm could result and that the employer knew, or should have known, of the hazard. A mandatory penalty of up to $7,000 for each violation is proposed. A penalty for a serious violation may be adjusted downward, based on the employer's good faith, history of previous violations, gravity of the alleged violation, and size of business.
- **Willful violation.** A violation that the employer knowingly commits or commits with plain indifference to the law. The employer either knows that what he or she is doing constitutes a violation or is aware that a hazardous condition existed and made no reasonable effort to eliminate it. Penalties of up to $70,000 may be proposed for each willful violation, with a minimum penalty of $5,000 for each violation. A proposed penalty for a willful violation may be adjusted downward, depending on the size of the business and its history of previous violations. Usually, no credit is given for good faith. If an employer is convicted of a willful violation of a standard that has resulted in the death of an employee, the offense is punishable by a court-imposed fine or by imprisonment for up to 6 months or both. A fine of up to $250,000 for an individual, or $500,000 for a corporation, may be imposed for a criminal conviction.
- **Repeated violation.** A violation of any standard, regulation, rule, or order where, upon reinspection, a substantially similar violation can bring a fine of up to $70,000 for each such violation. To be the basis of a repeated citation, the original citation must be final; a citation under contest may not serve as the basis for a subsequent repeated citation.
- **Failure to abate prior violation.** Failure to abate a prior violation may bring a civil penalty of up to $7,000 for each day the violation continues beyond the prescribed abatement date.

The penalties for poor safety practices at a construction site can be substantial. These numbers are for single violations. Several violations may be found in a single inspection. An example highway project where substantial penalties were given occurred on Route 3 in Massachusetts in 2004 (OSHA, 2004). It is discussed in the following press release.

Case Study: Safety Hazards on Route 3 Highway Project Lead to $371,000 in OSHA Fines for Massachusetts Contractor

Methuen, Mass.—The U.S. Labor Department's Occupational Safety and Health Administration (OSHA) has cited Modern Continental Construction Co., the lead contractor on a highway improvement project through Middlesex County, Mass., for failing to protect its workers against cave-in, fall and drowning hazards at four different worksites on Route 3 in Bedford, Billerica and Chelmsford. The company faces a total of $371,000 in proposed fines for the violations.

"To ensure that injury and illness rates continue to decline, we must make sure that employers protect employees from workplace hazards," said U.S. Secretary of Labor Elaine L. Chao. "The significant penalty of $371,000 in this case demonstrates the Administration's commitment to protecting the health and safety of American workers."

OSHA's inspection found employees at the two Chelmsford worksites exposed to cave-in hazards while working in unprotected excavations that also lacked safe means of escape. At both sites, the supervisors with the knowledge to spot the cave-in hazards and the authority to correct them failed to do so. In addition, workers at one site were exposed to falls of up to 28 feet from an unguarded walkway while workers at the other site faced crushing hazards from a crane that had been set on unstable ground and had not been inspected for defects.

Workers at the Bedford and Billerica sites who were required to work over or near the Concord and Shawsheen rivers faced serious injury if they fell in the water since the required life-saving skiffs, life vests and ring buoys were not available. At the Bedford jobsite, an unsafe raft used to transport workers across the Shawsheen River and an un-inspected and improperly positioned scaffold posed additional hazards.

Modern Continental was cited for four alleged willful, five alleged repeat and seven alleged serious violations of the Occupational Safety and Health Act. The willful citations account for $235,000 in proposed fines, $120,000 in penalties stemmed from the repeat citations and $16,000 in fines from the serious citations . . .

Cambridge-based Modern Continental has 15 business days from receipt of its citations and proposed penalties to request and participate in an informal conference with the OSHA area director or to contest them before the independent Occupational Safety and Health Review Commission. OSHA's Methuen area office conducted the inspection.

OSHA conducted almost 40,000 inspections in fiscal year 2003, an increase of 2,000 inspections over 2002 levels; most were focused on high hazard industries. The Occupational Safety and Health Administration is dedicated to saving lives, preventing injuries and illnesses, and protecting America's workers. (OSHA, 2004)

It has already been shown that an improved safety record can significantly decrease the number of costly accidents and OSHA fines. To implement safety improvement, most construction companies employ a safety management program. On small projects, the person responsible for the program may be the superintendent or project manager. For large projects, a safety engineer may be hired to manage the safety program and continuously identify any potential safety deficiencies. A safety engineer can be defined as a person who inspects all possible danger areas of the construction site for safety hazards, as well as defining a safety plan and overseeing on-site training of workers in safety issues.

Safety Management Plan

A safety management plan consists of four primary components (Jackson, 2004):

1. **Commitment from top management.** There must be a commitment from the construction company's top management. The most successful safety plans include the involvement and support of the top level of management of the construction company and must include employee involvement in the implementation of the plan.
2. **Safety analysis.** A safety management plan must include procedures to analyze the construction project to identify existing safety hazards and conditions that might become safety hazards. Workers should be trained to report hazards.
3. **Safety procedures.** A safety management plan must establish safety procedures to correct or control hazards on the job site. Issues addressed in this part of the plan include procedures for wearing personal protection, accident prevention procedures, and the provision of information about hazardous materials used on the project.
4. **Training.** Training is an important component of improving workplace safety. The construction contractor must make appropriate training available to all workers. The amount of training necessary is dependent on the size and complexity of the project, as well as the potential hazards. For example, roofers have a high rate of falling accidents. They would appropriately be given fall prevention training.

Safety Activities and Training in Construction

Several activities occur simultaneously on a construction project to improve safety performance. Provision of the proper safety equipment to individual workers is important. For example, if workers are to operate in a dusty environment, they must be provided with suitable respirators. How elaborate the respirators must be depends on the nature and toxicity of the source of the dust. Categories of protection that must be addressed by the contractor include body protection, head protection, eye protection, hearing protection, and fall protection. Hard hats are an example of head protection and protect the worker from falling objects.

Training is ongoing during most construction projects. For complicated tasks, workers might be given training classes in how to safely perform the task. More routinely, "tool box" meetings are held. They are led by a foreman or superintendent to inform a crew of safety problems occurring on the project. The meeting gives workers information on procedures to remediate the problems that are occurring on the project. For example, a meeting on a project involving considerable excavation and trenching could be called to discuss trench safety. The meetings are usually short and are typically held on a weekly basis.

Data are also kept on the construction site about the chemical make-up and dangers in using the various chemical products that are used on a project. Safety management plans contain a mechanism for logging these materials and making the information about the chemical product available to workers. An official document, the material safety data sheet, is provided by manufacturers for each product they make. Each data sheet contains information about the hazards associated with each product and recommended methods for handling and storage. Failure to maintain these records is citable by OSHA (Jackson, 2004).

Case Study: Toward a Culture of Safety

The following article describes how owners and contractors are working together in the construction industry to improve safety. This article, titled "More Companies Bring Jobsite Safety Up Close And Personal: Owners and Contractors Seek to Reduce Injuries by Including All Workers in Jobsite Safety Cultures," was published in *ENR* magazine on June 6, 2005. Debra K. Rubin wrote the article.

In the early 1990s, an Intel Corp. construction project in Ireland received a safety flag, a top local award for an outstanding jobsite safety record. But project leader Art Stout, now the chipmaker's manager of corporate capital development, wouldn't allow the flag to be flown, claiming the record "wasn't good enough."

Intel implemented and honed a tougher corporate-wide construction safety program that has made it an industry role model today. Recordable injuries and illnesses fell from a rate of 5.95 per 100 full-time workers in 1994 to 0.68 just five years later. But the global manufacturer, which spends between $3 billion and $7.5 billion a year on capital investment, still is not satisfied. "We used to think injuries were a natural course of construction," says Stout. "Now we want to get our injury rate beyond zero."

Reaching that goal has pushed Intel to look at safety in a new light and to elevate its jobsite priority to a new level, even above schedule and budget. "We're willing to sacrifice those two to make sure no one gets hurt," says Stout. For Intel and other owners and contractors, safety these days is permeating the very fabric of the jobsite—creating a "culture" that demands buy-in from top executives and workers alike, and transforming safety management from simply collecting statistics to predicting why deaths and injuries happen and proactively preventing them.

Despite improved safety performance on construction sites over the years and more exchange of best practices among owners, organized labor, union and nonunion contractors, accidents such as the March 23 explosion at a BP refinery in Texas that killed 15 contractor employees in trailers remind industry officials that all is far from perfect (ENR 4/4, p. 10).

Oil company officials have taken responsibility for the accident and are making reparations, but talk abounds of breakdowns in site safety communications and of contractors hurriedly removing trailers from the refinery boundaries. "In a long-term sense, it's scary," says Ron Prichard, a Plainfield, Ind., safety consultant also affiliated with the Construction Users Roundtable (CURT), a Cincinnati-based group of large owners. "For the last 15 years, everyone has been losing staff in safety and we've operated lean for so long, we're just a cat's whisker away from a disaster again sometime soon."

Intel and other enlightened owners are spreading the safety responsibility to everyone on site and deep into the company fabric. "When I arrived at Intel three years ago, I got a cup of coffee and an employee warned me to put a lid on the cup," says Bob Predmore, director of worldwide construction. "It may sound simple, but it embeds the culture."

Intel contractors know from the start that the owner means business on safety. Construction bidders must have an "experience modifier rate" of 1.0 or lower, a figure that governs workers' compensation insurance costs and signifies a responsible firm. "We're now dealing with contractors that have .6, .7, and .8 three-year rolling averages," says Steve Bowers, Intel's worldwide safety director. Contractors feel the pain but know it boosts their credentials. "They will grade you on your safety record," says one Intel builder. "They force you to put your money where your mouth is."

Intel officials believe that injuries are not a cost of doing business. "It's all about leadership and the time and money it takes to get to that point," says Bowers. The chipmaker has led the industry in studying soft tissue injuries, a big problem in construction, says Ali Afghan, construction program manager at Intel's Oregon site. "Our vision is to raise the bar to another level and see the reduction of muscular and skeletal cumulative trauma disorders." Intel contractors now must define an ergonomics program to cut soft tissue injuries.

Shutting down a project in the cutthroat semiconductor market, even if for just a "near-miss" incident, amplifies Intel's safety culture message, officials insist. "We stopped a project for two days that was already two weeks behind the critical path," says Predmore.

Afghan admits that the safety credo is a business decision, preserving long-term work force mobility and quality of life. Intel also notes major improvements at overseas sites, particularly in cultures where safety rules never existed or were eschewed. Workers at one Chinese fab plant site once rioted over safety goggle requirements, says Predmore. Last month, he cited the job for 3.5 million man-hours without a recordable incident.

Officials at Chevron Corp. credit development of the oil company's "incident and injury-free" safety culture to a small, Austin, Texas-based consulting firm that helped it turn around results at a $650-million fuels plant and pipeline in Saudi Arabia in 1996 . . .

"Our safety program had been unpredictable," says Rick Miller, a Chevron project manager who hired JMJ Associates. "In one year, we had 13.5 million man-hours with one lost-time accident and two recordables." After such results, Chevron tried the JMJ approach at a few pilot sites. Safety statistics were so immediately impressive that the owner used it at all large capital construction jobs. The firm now is using the method for operations and maintenance at refineries.

The JMJ approach builds on a company's safety processes and procedures by adding a new layer of commitment, says Steve Knisely, a JMJ partner. It requires shifting safety from a priority to a "value," a deep-seated belief that it will not be compromised and will actually drive a company's actions. "We have a process, but the company does the work," says Knisely, who also counts Intel as a client. "It brings the human element to the workplace and creates an approach for people to understand the consequences of injuries and death." Adds JMJ founder Jay Greenspan, "The construction industry is statistically obsessed but numbers come down when you focus on people."

The approach focuses on safety's "subjective side," adds Craig May, a Chevron capital project coordinator. "It studies what can't be measured—behavior, culture and intentions." Chevron's safety program now gives equal emphasis to subjective issues and processes, he says. May was not always a believer, fearing the approach would be considered too warm and fuzzy for a hard-edged jobsite. "At first, I wondered if we were going to sit around and sing Kumbaya," he says.

On May's latest project, the hookup and commissioning of Chevron's Sanha Condensate complex off the coast of Angola that finished this year, 1 million man-hours of work were completed with no lost-time injuries and only one recordable injury. For all Chevron projects, recordable injury rates are down 50% and total days away from work have dropped 60%, officials say. May says top management supports the new approach.

Contractors also are required to sign an agreement with JMJ to use its services but getting their buy-in is not always easy. On Chevron's oil and gas development project in Tengiz, Kazakhstan, a joint venture of WorleyParsons and Fluor Daniel that held the project's EPC contract, balked at the requirement. "It took an effort to compel them," says Gary Fischer, Chevron project coordinator. Fluor officials confirm the initial skepticism, citing the firm's already good safety record. But with evidence of early results, Fluor embraced the approach and now is using it on projects run by three of its large offices, says a spokesman.

Public Problem

With procurement rules tougher to manage in public sector construction, owner-led safety cultures are fewer, officials say. "It would be desirable to include a safety record in the contractor selection process, but we'd have to figure out how to measure it and how to write it into contract language," says Bob Hixon, a former top construction official at the U.S. General Services Administration, who now manages the $540-million U.S. Capitol Visitors Center.

Even so, some public owners are taking on the task. At the $1.5-billion upgrade and expansion of the University of Cincinnati, officials are making safety a more visible and personal priority for contractors, site workers, community stakeholders and students.

"We went from doing a lot of construction with no safety program to winning awards," says Raymond D. Renner, director of construction management. The proximity of 35,000 students and 14,000 staff members propelled the change. The university brought on Joe Glassmeyer, project manager at Cincinnati-based Messer Construction to manage safety activities.

Among the program's features is a 40-page safety "manual" that spells out the site's environmental, safety and health requirements and must be signed by all site participants. Contractors initially wanted a change order for the time it took, but then decided it was helping site productivity, says Glassmeyer. "We're getting better contractors, fewer incidents and no accidents involving students, faculty or the public," he adds. "We get buy-in from workers and teach them how to act in our home on campus."

This month, the university is among a group of local owners funding a new drug-testing program. Obtaining the drug certification enables workers to take jobs on projects of all participating owners. "We thought there would be grumbling or protests by unions, but no one complained," says Glassmeyer. "It's something we had to do."

Building trade unions are working with CURT to promote having safety and health criteria included in bid document and selecting contractors based on their safety and health records instead of low bids, says Pete Stafford, executive director of the Center to Protect Workers' Rights and director of safety and health for the AFL-CIO's Building and Construction Trades Dept. The unions also want mandatory safety training required in project labor agreements.

Colorado Springs Utilities prequalifies all of its contractors based on safety plans and past records. "We have disqualified certain contractors for not meeting our standards," says Chief Operating Officer Jerry Forte. The approach accompanied him to Colorado Springs when he left the U.S. Dept. of Energy's Los Alamos National Laboratory, where site construction contractor Johnson Controls had implemented it. "It didn't need me to sustain it. It was embedded in everyone's heart and soul," he says.

Johnson Controls shifted the culture at the DOE site in response to safety issues, says Jon M. Barr, retired president and general manager of the firm's New Mexico-based unit. "We needed to take a softer approach," says Barr, a retired Navy admiral.

Contractors also are looking beyond lagging indicators of reported incident numbers, focusing instead on leading indicators, such as leadership and employee behavior. "Those are very subjective reviews, but that's where we can intervene and stop an accident rather than simply reacting to numbers," says John Mathis, manager of environment, safety & health for Bechtel Systems & Infrastructure.

Skanska Building, Parsippany, N.J., has implemented a number of JMJ innovations to build its jobsite safety culture and improve safety results. "In construction, we're all right-brained engineers. If people get hurt, you just give them goggles and hardhats," says Paul Anderson, COO of its Southeast region. "If that doesn't work, you just give them better goggles and harder hardhats." He stresses that discussion of "near-misses" on site is equally important. "Instead of safety staffers policing jobsite people, people on the job have to police each other."

Skanska has started a prework exercise program at its sites to reduce injuries and now is seeking cost allowances in its contracts and talking to unions and subs. Anderson notes that on a $45-million Austin museum project, a site bulletin board features photos of workers' families. There has been just one injury.

Birmingham, Ala.-based BE&K learned its safety values from DuPont, an industry pioneer, says Ted Kennedy, one of the contractor's founders. "We have a responsibility to see that everyone leaves our job safely," he says. Employees have to know that finishing a job now is not more important than their safety, says Rich Baldwin, safety director. "They can never let a super hurry them," he says. "They have the right to make the job safe."

Summary

This chapter has shown that accidents are costly and can reduce a contractor's competitiveness. There are direct accident costs related to the amount of insurance premiums a contractor must pay for a poor safety rating. Accidents have other indirect costs that are not reimbursed by insurance. OSHA is an agency of the federal government that regulates job-site safety in the United States. OSHA can inspect a construction site at anytime and may levy substantial fines for unsafe conditions. To reduce accidents, contractors typically maintain a safety management plan. An important component of the plan is to provide workers with training classes and meetings about safety problems. On large projects, a safety engineer may be used to identify and mitigate hazards as they occur on the project.

Key Terms

Experience modification rating (EMR)

Occupational Safety and Health Administration (OSHA)

Self-insure

Workers compensation insurance

Review Questions

1. What type of accident causes the most fatalities in the U.S. construction industry?
2. Discuss the difference between direct and indirect accident costs.

3. Why does a construction company with a low EMR have a bidding advantage over a company with a high EMR?
4. A willful violation of OSHA regulations can result in a fine as large as:
   - $75
   - $7,000
   - $700
   - $70,000
5. What are the components of a construction company safety program?
6. What is a "tool box" safety meeting? How does it differ from the safety courses that a worker performing a complicated task like welding receives?

Management Pro

MANAGEMENT PRO

Observe a construction project. Note the safety precautions taken. Is proper safety equipment used? Do you observe safe working conditions and practices? Check OSHA regulations to determine if the work conforms to the OSHA requirements. Write a short report describing what you observed.

References

Bureau of Labor Statistics. 2008a. Census of fatal occupational injuries (CFOI). Current and revised data, Industry by event or exposure, 2007. Available from www.bls.gov/iif/oschfoi1.htm#2007 (accessed September 20, 2008).

Bureau of Labor Statistics. 2008b. Injuries, illnesses and fatalities, Numbers of nonfatal occupational injuries and illnesses by industry and case type, 2006. Available from www.bls.gov/iif/oshsum.htm (accessed September 23, 2008).

Engineering News Record. 1991. Jobsite safety pays off in insurance savings. *ENR* vol. 226: 23.

Halpin, D. and R. Woodhead. 1998. *Construction Management*. New York: Wiley.

Hinze, Jimmie W. 1997. *Construction Safety*. Upper Saddle River, NJ: Prentice-Hall.

Jackson, Barbara J. 2004. *Construction Management Jump Start*. Alameda, CA: Sybex.

Levitt, Raymond E. and Nancy M. Samuelson. 1993. *Construction Safety Management*, 2nd ed. New York: John Wiley and Sons.

OSHA. 2004. Safety hazards on route 3 highway project lead to $371,000, OSHA fines for Massachusetts contractor. OSHA regional news release. Available at http://www.osha.gov/oshaweb/owadisp.show_document?p_table=News_Release&P_id=10776 (accessed September 27, 2008).

Princeton University. 2007. Introduction to OSHA. Available at http://web.princeton.edu/sites/ehs/healthsafetyguide/f1.htm (accessed September 17, 2008). Section F1, Health and Safety Guide.

Reason, James. 1997. Managing the Risks of Organizational Accidents. Aldershot, UK: Ashgate.

Safety Management Group. 2002. The experience modification rate (EMR). Available at http://www.SafetyManagementGroup.com/articles/The-Experience-Modification-Rate-(EMR).aspx (accessed September 22, 2008).

U.S. Department of Labor. 1996. OSH Act. OSHA standards, inspections, citations and penalties. Available at http://www.osha.gov/doc/outreachtraining/htmlfiles/introsha.html (accessed September 15, 2008).

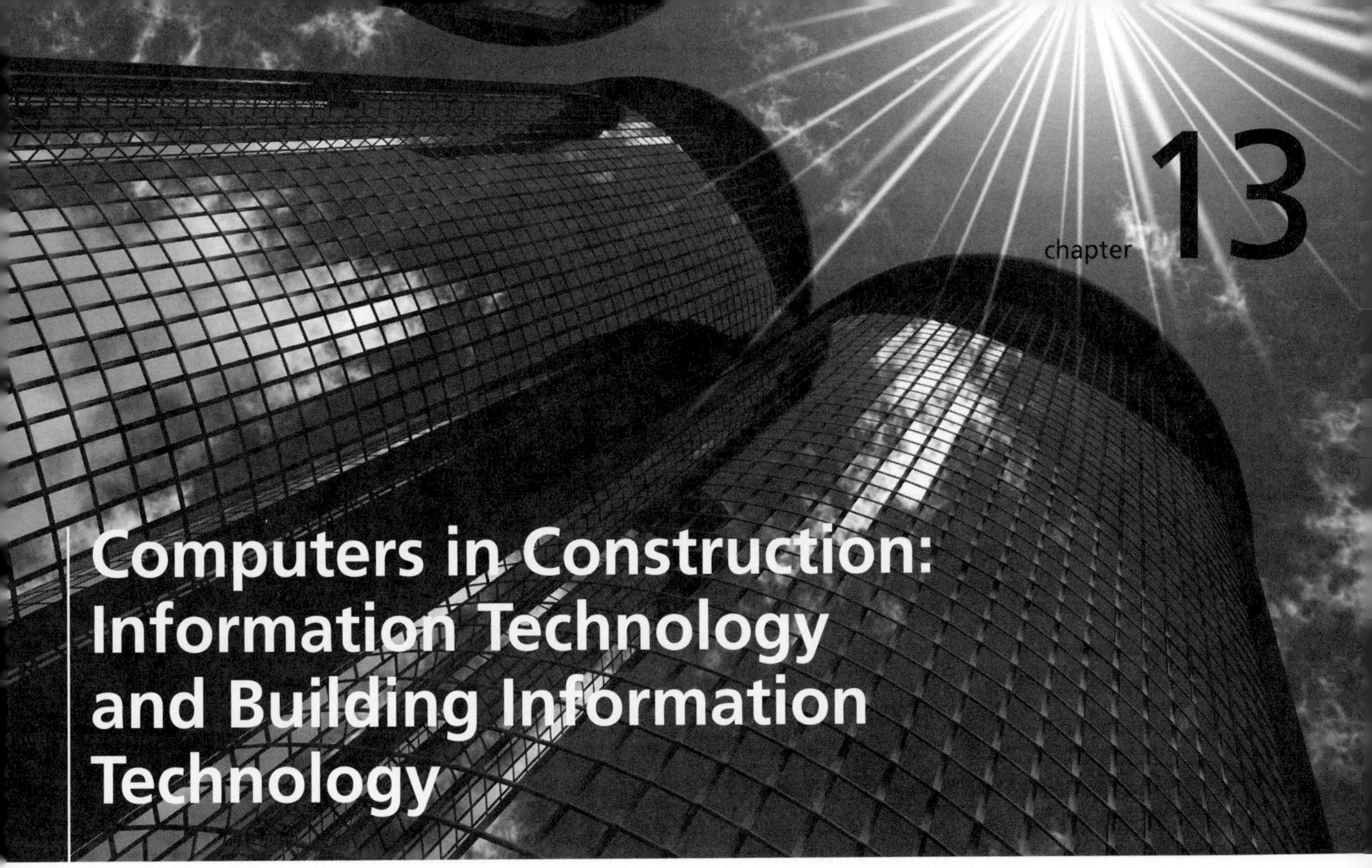

chapter 13

Computers in Construction: Information Technology and Building Information Technology

Chapter Outline

Introduction

Building Information Modeling

Information Technology for Construction Management

Wireless Networks

Web-Based Applications

Construction Applications of Mobile and Wireless Computing

Summary

Key Terms

Review Questions

References

Introduction

Information technology (IT) is a term that encompasses all forms of technologies (computers, software, and telecommunications) used to create, store, exchange, and use information in its various forms (business data, voice conversations, still images, motion pictures, and multimedia presentations). Advances in IT have made it increasingly possible to apply advanced computer techniques to the design and construction of buildings. Many technologies that are now commercially feasible have emerged only in the last few years. A construction manager must keep abreast of IT developments to insure that his or her firm keeps its competitive advantage by using the latest advances in IT to increase productivity.

This chapter will focus on two of the most exciting areas of IT development and application in the construction industry:

- The use of building information models, which are powerful computer models that aid in the visualization of complex projects for design and construction
- The use of wireless communications, the Internet, and portable computers in the field to radically change how data and information flow on construction projects

Building Information Modeling

Building information modeling (BIM) centers on the use of virtual models (typically 3D or 4D models of a facility), which allows information and data to be linked to objects in the virtual model. This in turn allows data such as cost and schedule to be linked to the objects in the drawing, allowing a single model to represent the physical attributes of the facility as well as all the other important information required to construct the facility.

Introduction to BIM

Two-dimensional models (such as traditional construction plans) require multiple views to depict a 3D object in adequate detail for construction. Additionally, traditional paper plans and CAD drawing files are stored as lines, arcs, and text, which are not computer readable and cannot be acted upon by computer programs other than CAD. In BIM, the design is represented as objects that carry their geometry and attributes. If an object is changed or moved, it need be acted upon only once, and all the various views needed during planning and construction can be generated. This differs from traditional CAD, where each separate 2D or 3D drawing would need to be changed, possibly introducing inconsistencies in the plans (Eastman, 2007).

Three-dimensional, four-dimensional, and BIM technologies represent three separate, but synergistic, ways in which computer technologies can aid in the management of facilities throughout a project's lifecycle. 3D geometric models are the geometric representation of building components and typically serve as an aid for visualization and design/construction coordination (Finith, 2007). A 3D model on its own does not amount to BIM. 4D models (3D + time) include information that can inform and analyze project phasing, tenant sequencing, and construction scheduling. Using a 4D model, it is, for example, possible to view in 3D how a building would evolve over time if constructed according to its CPM schedule. BIM includes not only 3D and 4D geometric models but also more specific data on a wide range of building elements and systems associated with a building (e.g., wall types, spaces, air handling units, geospatial information, and circulation zones) (General Services Administration, 2007).

The General Services Administration (2007) defines BIM as the development and use of a multifaceted computer software data model to not only document a building design but also simulate the construction and operation of a new or a rehabilitated facility. The resulting BIM is a data-rich, object-based, intelligent, and parametric digital representation of the facility, from which views appropriate to various users' needs can be extracted and analyzed to generate feedback and improve facility design.

Benefits and Uses of BIM

One benefit of the use of BIM is the capability to check for spatial conflicts. BIM can provide 3D plans and renderings. Three-dimensional plans can be used from the initiation of the design through construction. Three-dimensional CAD is employed by designers, including architects and engineers, to visualize designs and identify conflicts. Three-dimensional drawings can also be used during construction by construction contractors to interpret complex designs. The benefits of applying 3D drawings to the construction process include (Cory, 2001):

- Checking clearances and access
- Visualizing project details from different viewpoints
- Using the model as a reference during project meetings
- Enabling constructability reviews
- Reducing interference problems
- Reducing rework

Figure 13.1 shows the complex visualizations that can be generated using BIM. The figure shows various views of a complex building to enhance the understanding of construction details. Figure 13.2 shows how BIM can be used to generate cross sections at points of interest in a building.

Identification of conflicts has most extensively been applied to complex process plants, where it may be difficult to visualize how networks of pipes may conflict using 2D drawings. The use of 3D plans has naturally focused on process plants and complex

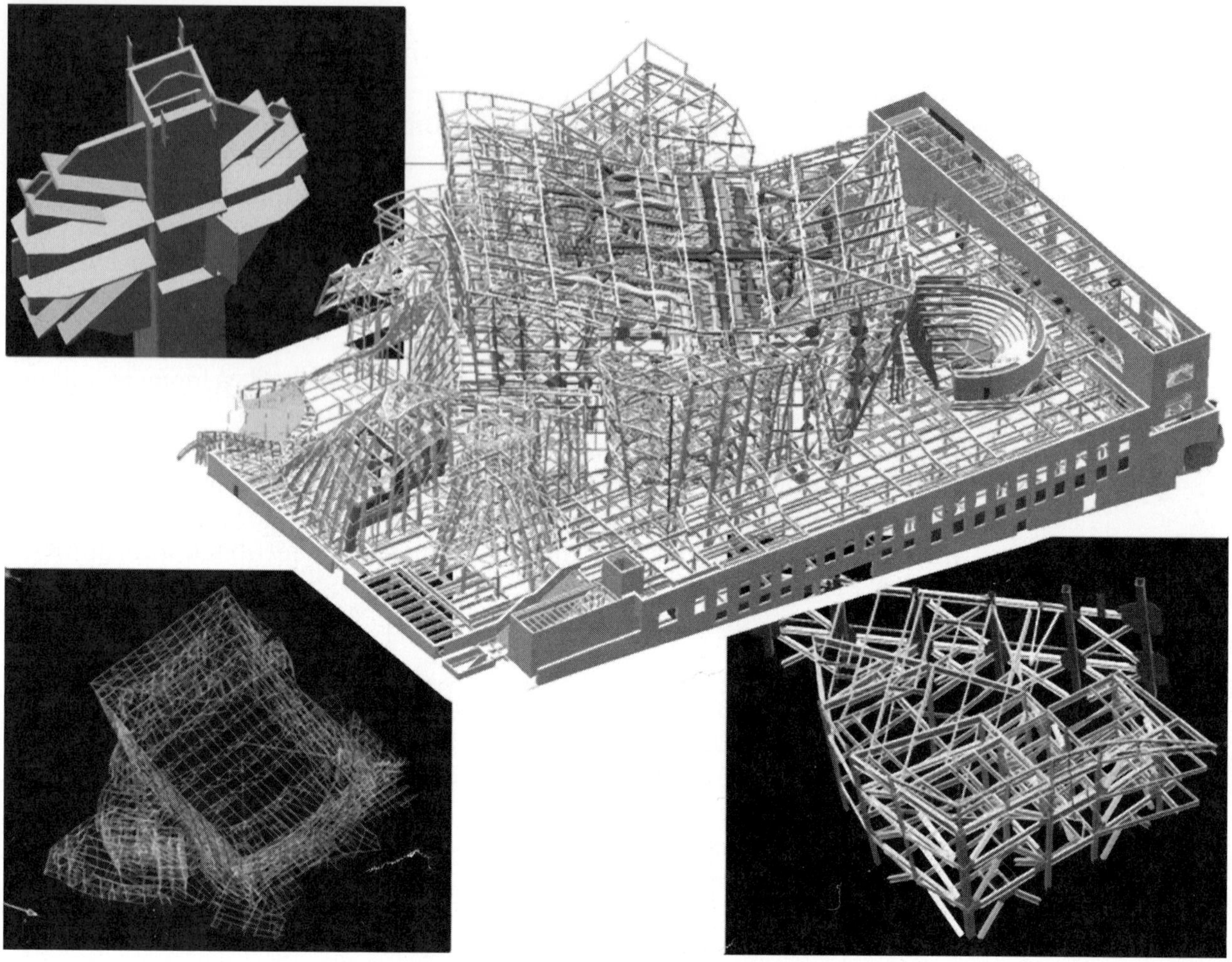

Figure 13.1 Visualizations from a complex building. *Images courtesy of Gehry Technologies.*

commercial buildings. Their use is now being extended to infrastructure design such as highways and railroad alignments.

Interoperability

One of the great benefits of BIM is that it can integrate different types of analyses together; that is, it can allow different types of computer programs to work together. Interoperability is a motivating factor for the application of BIM for design, construction, and operations of a constructed facility. Interoperability is the ability of a system or a software product to work with other systems or products without special effort on the part of the customer. BIM provides interoperability between various software programs without the need to manipulate or reenter data. A study by the National Institute of Standards and Technology (NIST) has estimated that the lack of

Figure 13.2 Three-dimensional cross sections produced by BIM. *Courtesy of VICO Software Inc.*

interoperability in the construction industry adds a total of $15.8 billion annually to the cost of construction projects (Gallaher et al., 2004). One of the significant findings of the NIST report is that a lack of data standards inhibits the transfer of data between different phases of a project's life cycle. BIM has the potential to reduce the cost of inefficiencies of transferring data and to allow for smoother data flow between different project participants and between different project phases.

BIM allows the building object data to be used in many ways, including generating automatic bills of material allowing for cost estimating and automatic ordering and tracking of materials. Many other applications are possible using the BIM data, including scheduling, lighting, water flow, and acoustic analysis. These functions can be performed during design rather than waiting until after the design is complete—when it would be costlier to modify the project.

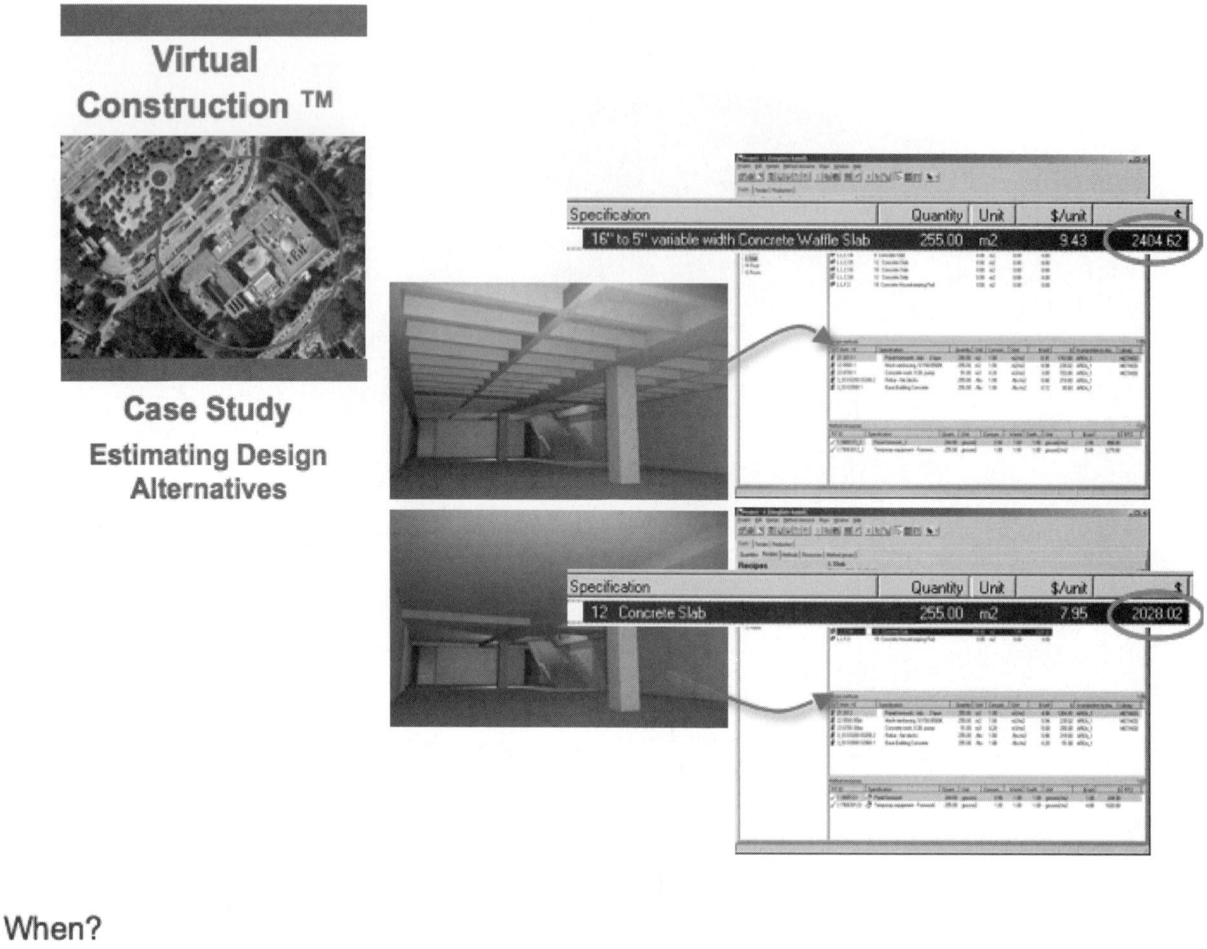

Figure 13.3 Evaluation of design alternatives. *Courtesy of VICO Software Inc.*

Figure 13.3 shows how BIM allows for design alternatives to be analyzed. The figure shows how 3D renderings of two ceiling designs can be analyzed and their costs compared. The data in BIM allow the data about the geometry of the building objects to be integrated with construction costs.

BIM can be used to automatically generate bills of materials and shop drawings. Shop drawings are detailed drawings that are sent to material fabricators to allow them to construct building components. In the past, quantities would have to be calculated by an estimator from 2D plans to develop a bill of materials, which allowed the materials needed for a project to be ordered well in advance. This process, using BIM, is shown in Figure 13.4.

In BIM, the schedule can be linked to the 3D building, creating a 4D view of the building. In 4D CAD, changes in a building during construction can be simulated. Typically a BIM system integrates a CPM scheduling software, and the building construction is simulated according to the activities in the CPM schedule. This is useful for planning

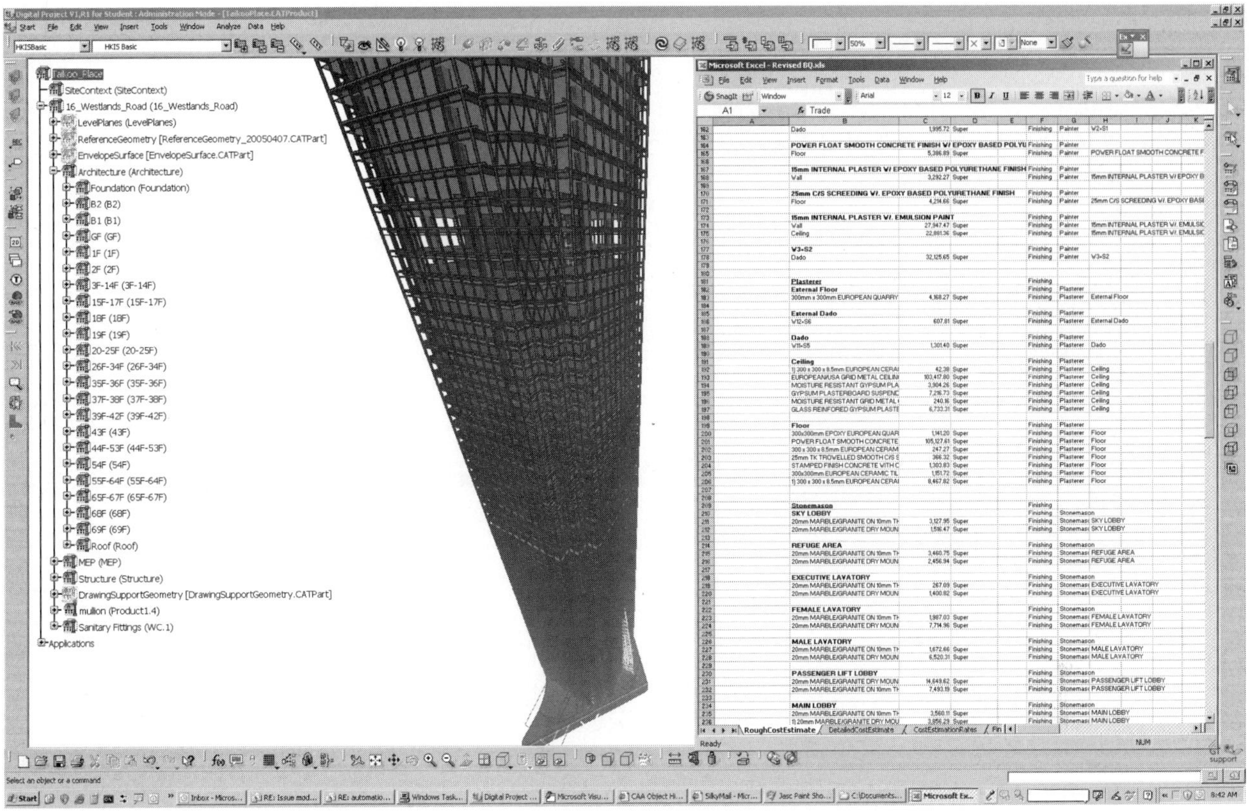

Figure 13.4 Bill of quantities extraction. *Courtesy of Gehry Technologies.*

purposes and identification of conflicts between crews and subcontractors. On one large stadium project, 4D CAD was used to determine where two large cranes could be located so they would not interfere with each other. Figure 13.5 illustrates a BIM program where the 3D CAD model is linked to a Microsoft Project schedule.

Some BIM Software Applications

There are different computer programs that have BIM features. The state of the art of this technology is changing rapidly. New products are continuing to arrive on the market. A good source of information about emerging trends in this area is the *Constructech* magazine (www.constructech.com). A brief summary of some BIM software programs is given here to illustrate what is possible when using BIM.

1. **Bentley Navigator.** Bentley Navigator is a program with the capability of linking 3D models with either Primavera or Microsoft Project. The software generates simulations of construction schedules and heavy lifts, providing an understanding of object motions and potential clashes in an area of activity (Sheppard, 2004). The schedule data is dynamically linked to the 3D drawings, and modifications to the schedule can be immediately visualized in the 3D environment.

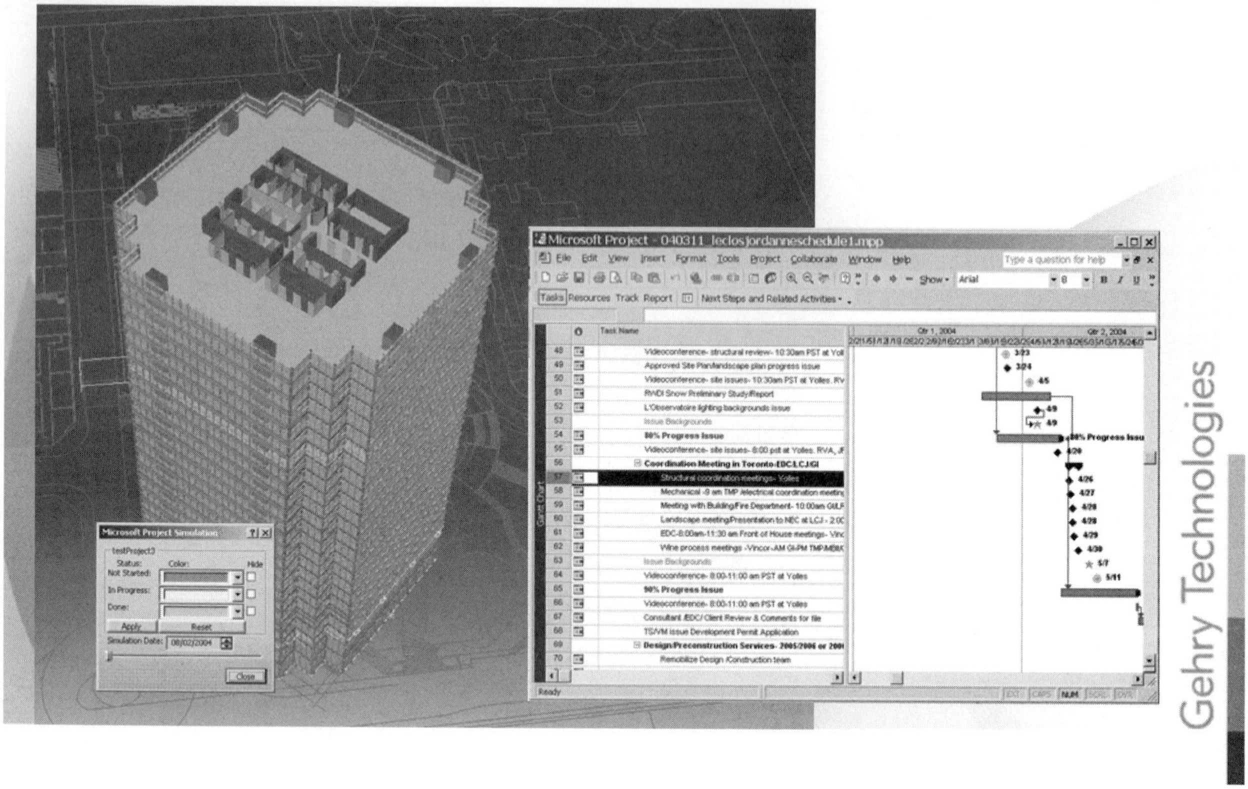

Figure 13.5 Construction schedule simulation using BIM. *Images courtesy of Gehry Technologies.*

2. **Balfour Technologies's fourDscape.** Balfour Technologies has developed a visualization modeling system called fourDscape. This software is server based and users access the program through a Web browser. The software allows users to interact with 3D models of time-dependent information. Users can step forward and backward in time to identify relationships between visualized 3D objects (Sheppard, 2004). Using the fourDscape program, dates are assigned to 3D objects. The fourDscape model does not include any sort of scheduling algorithm. Therefore, it would be necessary to carry out a CPM analysis to determine the dates to assign to the objects. However, the primary benefit of using this program is the capability to allow team members in different locations to view 4D visualizations using a Web browser (Heesom and Mahdjoubi, 2004).
3. **Gehry Technologies's Digital Project.** Gehry Technologies sells a program called Digital Project (Gehry Technologies, 2005). Digital Project is marketed as an advanced building information modeling and construction management system. Gehry Technologies has extensive experience in applying its software to the projects of noted architect Frank Gehry. Gehry's projects are known for their unusual designs and use of curved shapes, and are ideal candidates to be managed using BIM. The Digital Project software is based on the CATIA software developed

by Dassault Systemes, a French aerospace company. Some interesting features of Digital Project include:

- Bidirectional links between Digital Project and Microsoft Excel
- Compatibility with existing 2D and 3D file formats such as DWG and DXF
- Integration of project management functions with the 4D-building model; using Microsoft Office, linked templates for design issues, change orders, and requests for information can be produced
- Linking of the 3D model to Primavera scheduling software provides 4D navigation

4. **Common Point 4D.** Common Point 4D is another powerful 4D CAD program that can be used for both constructability analysis of a structure and layout and planning of a construction site. The program can accept scheduling data from both Primavera and Microsoft Project. It also accepts various types of CAD files as input. One interesting feature is the ability to accept a Tekkla Structures LE file, which is a program used to create 3D drawings of steel detailing.
5. **Graphisoft Constructor.** Graphisoft Constructor is a 5D model that provides an integrated solution for project visualization, scheduling, and estimating. The basic Constructor program provides 4D modeling using integration with Primavera. Additional modules of the Constructor program are Estimator and Control. When using Estimator, every object in the 3D model is connected to an estimating recipe. In turn, this recipe information can be used to define schedule durations. The Estimator program includes the capability to perform Monte Carlo simulations where individual activities can be defined as statistical distributions. Cost variance reports can be produced that show project elements that may be prone to cost increases. The Estimator program is linked to the 3D building model and can also be used to rapidly assess the cost impacts of design alternatives. The Control software is an interesting feature of the Graphisoft Constructor offerings because it provides an alternative paradigm for scheduling based on location. The Control software generates a line of balance diagram for a selected location in a building. The line of balance diagram explicitly shows where conflicts occur between different resources at the selected location. The line of balance schedules are created from Estimator recipes and 3D model quantity information. Tasks can be defined by grouping materials and defining a production factor. Figure 13.6 shows a line of balance schedule created in Control linked to a 4D view of the project. Schedules created can be imported into the Constructor 5D model and simulated using the Control software. (Gallelo et al.)

There are several BIM models that focus on plant construction and maintenance. These programs are useful not just for the original construction but are designed to be used throughout the total life cycle of the plant to handle maintenance and retrofits. These programs serve the additional role of a database of the plant's components and layout. They represent a method of documenting and visualizing modifications, for example, to a complex industrial or chemical plant. Such programs include:

- Intergraph SmartPlant Review (developed by Intergraph)
- Plant CMS/Database (developed by Computer Systems Associates)

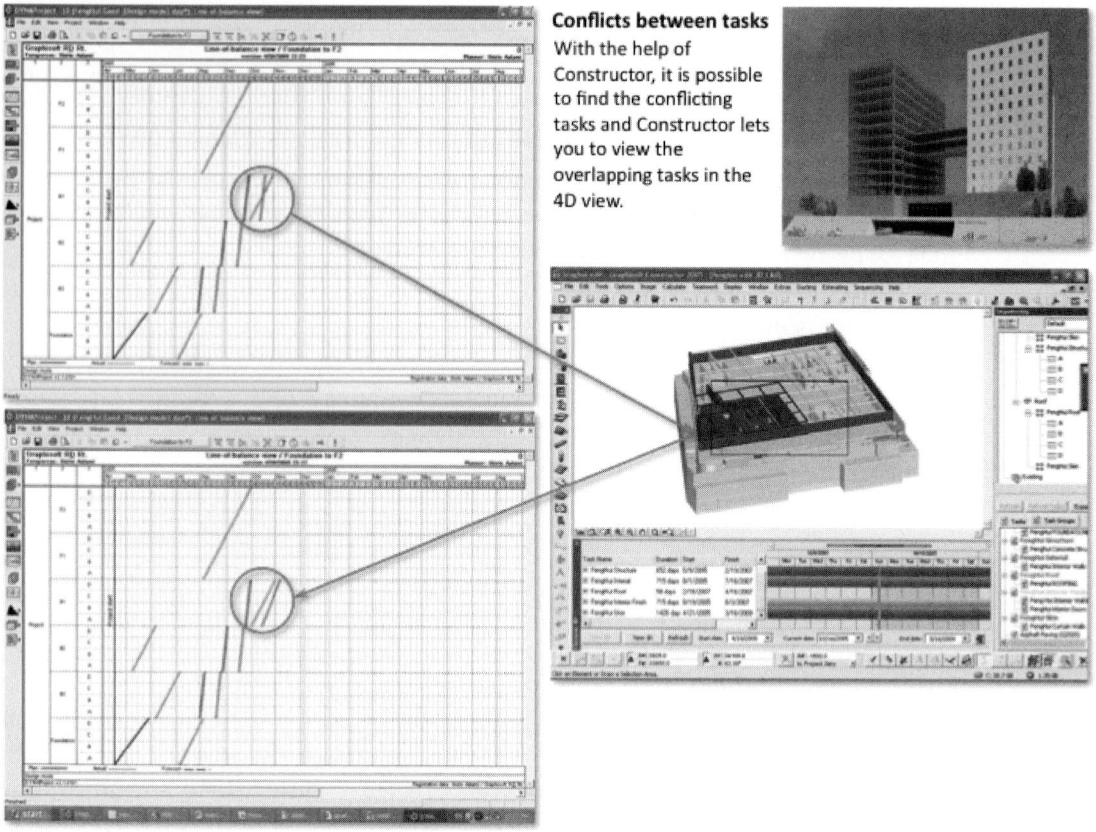
Conflicts between tasks
With the help of Constructor, it is possible to find the conflicting tasks and Constructor lets you to view the overlapping tasks in the 4D view.

Figure 13.6 BIM with line-of-balance capabilities to detect scheduling conflicts. *Images courtesy of Vico Software Inc.*

This review of the software is not intended to be comprehensive. New software are continuously being developed. This review of some of the best-known software indicates that there are differences in approach and features between the different programs. 4D software must be carefully evaluated before purchase to insure that its features are the best match for the types of projects a firm conducts.

Coordination and Collaboration Using BIM

It can be seen from the examples in preceding text that the use of BIM in a project promotes coordination and collaboration throughout the project life cycle. A single model is developed that can be used by all the members of the project team. The owner, designer, and contractor can use BIM at all stages of the project. For example, BIM is developed by the designer but can be used as a method of collaboration with the contractor to discuss constructability issues at an earlier stage in the project. At the end of the project, BIM can continue to be used by the building's facilities managers to control modifications to the building over time.

Information Technology for Construction Management

In this section, some basics of IT will be discussed and the ramifications of the new technologies on construction practice will be examined.

Basics of the Internet and Client/Server Computing

A construction manager must understand the basics of computers and computer networks because future practice will require managers to make purchasing decisions concerning complex computer systems and communications technology. Most readers of this book have probably used a Web browser to get information from the World Wide Web (WWW) over the Internet. However, it is appropriate to provide a clear definition of what the Internet and the WWW are. The Internet is the collection of public computer networks that are interconnected across the world by telecommunications links. It comprises millions of computers world wide, including those in businesses, government agencies, universities, and those of personal users. It allows users to communicate with each other using software interfaces such as electronic mail, FTP, Telnet, and the WWW. The WWW is the part of the Internet where connections are established between computers containing hypertext and hypermedia materials. The materials on the WWW are accessed through a WWW browser (such as Internet Explorer, Firefox, or Safari) that provides universal access to the materials that are available on the WWW and the Internet.

Computer networks are the way computers are linked together. Networks are widely used in the construction industry, from a small office network to large networks that tie together the geographically diverse offices of an international contractor. Many of the most popular construction software programs can now be networked in a LAN and/or accessed through the Internet. Construction managers must know the capabilities of these technologies because they may be called upon to buy or specify equipment for their projects. A basic definition of **client/server computing** is that it divides a computer application into three components: a server computer, client computers, and a network that connects the server computer to those of the clients (Lowe, 1999).

One definition for a server is a computer that delivers information and software to other computers linked by a network. Another definition for a server is a computer that handles requests for data, e-mail, file transfers, and other network services from other computers (i.e., clients). Personal computers (PCs) are typically used as clients. The equipment used for servers are more varied. A server can be any computer, from a PC to a very large array of computers or a mainframe computer. The selection of the server will vary depending on the application and the number of users.

Types of Networks

There are several types of data networks where computers are connected by wired lines. Two of them are:

- **Local area networks (LAN).** A group of computers and other computer devices such as printers and scanners that are connected within a limited geographic area.

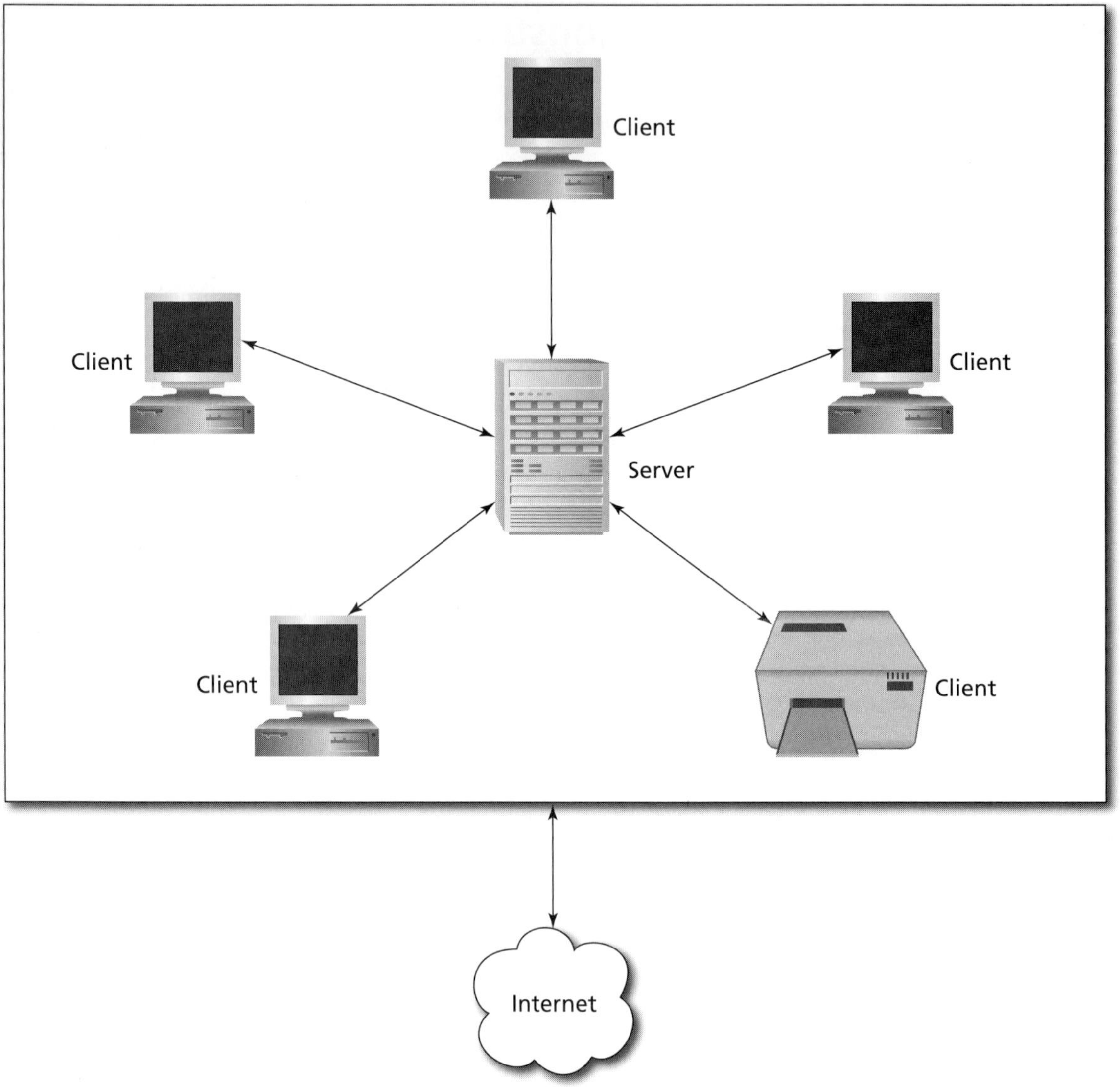

Figure 13.7 Typical local area network. From WILLIAMS. *Information Technology for Construction Managers, Architects, and Engineers* 1E. © 2007 Delmar Learning, a part of Cengage Learning Inc. Reproduced by permission.

LANs represent the type of network used to connect computers together in an office or project trailer (Dodd, 2005). Figure 13.7 shows the basics of a LAN. LANs are configured to allow users access to the local network and access to the Internet.

- **Wide area networks (WAN).** A group of LANs that communicate with each other. WANs are employed for connecting LANs that are widely separated geographically. WANs are commonly employed by large construction companies to connect their offices together. The LANs that compose the WAN can be connected together in a

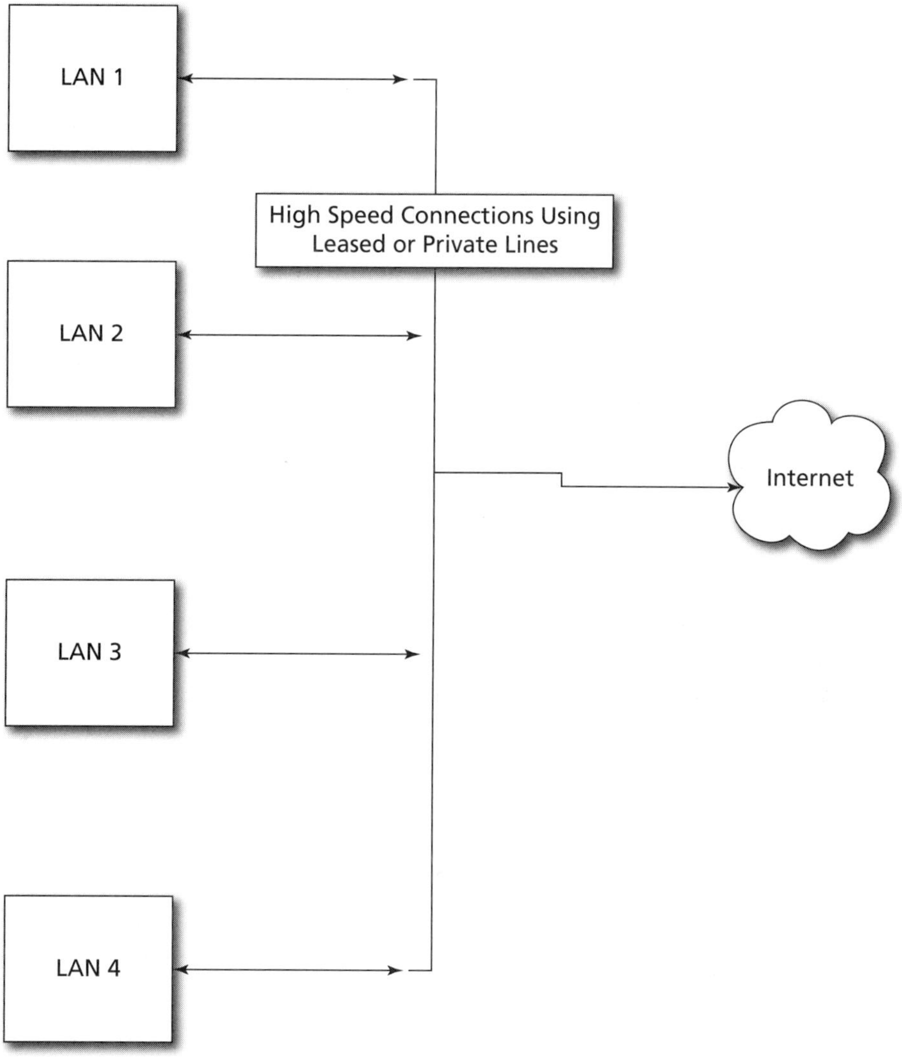

Figure 13.8 Typical wide area network. From WILLIAMS. *Information Technology for Construction Managers, Architects, and Engineers* 1E. © 2007 Delmar Learning a part of Cengage Learning Inc. Reproduced by permission.

variety of ways, including connection to the Internet and private high speed data connections. A basic WAN configuration is shown in Figure 13.8.

Implementation of client/server computing using a LAN or a WAN requires a network operating system software to be installed on the server. The client PCs must also have a compatible operating system software installed. There are many well-known providers of server software. These include Novell, Sun, Microsoft, and Hewlett-Packard. There are also open source server software that are widely used. The type of software used for the server software can be dictated by the requirements of the application software you want to run on your network.

Wireless Networks

A wireless network uses radio frequency technology to transmit and receive data. Implementation of wireless networks at the construction site requires some understanding of the types of networks available and their capabilities. The development for wireless networks has many potential applications in the construction industry. Wireless networking coupled with mobile computers can provide people in the field with access to a wide variety of programs and Web-based software. The advent of wireless network frees managers and workers from the need to go to an office to use the computer. They can spend more time in the field managing the construction process.

Wi-Fi Networks

The most common type of wireless local area network (WLAN) is the Wi-Fi network. Wi-Fi stands for wireless fidelity. When dealing with Wi-Fi equipment, a common term is 802.11. 802.11; it refers to the family of IEEE standards for building Wi-Fi networks. Most Wi-Fi equipment that is manufactured conforms to these standards. The 802.11 standards are evolving over time, with the main standards currently being 802.11a, 802.11b, and 802.11g. 802.11g offers higher speeds than 802.11b. 802.11a networks have a shorter range but are less prone to interference from other electronic devices. Both 802.11g and 802.11b have ranges of about 300 feet but the range can be extended at a construction project using devices called repeaters (Sandsmark, 2004). A new standard, 802.11n, is currently being developed. It will offer faster data speeds and longer range (Dodd, 2005).

Figure 13.9 shows a configuration of how Wi-Fi equipment might be employed on a construction project. Depending on the configuration of the construction site, several access points will probably be needed to provide full coverage of the construction site. The configured network can be a mix of wireless and cabled network or be entirely wireless.

Wireless Networking technologies are continuing to evolve. Technology that provides wireless networking over a broad area is evolving. Many cities are beginning to deploy Wi-Fi networks that provide coverage throughout a metropolitan area. These developments indicate that access to Wi-Fi networking for construction contractors will be easier in the future.

Web-Based Applications

Web-based applications are computer programs that run on a Web browser. These applications do not require any software to be installed on a user's computer. The computer software and all the data are stored on the server that the user is connected to. This provides several major advantages. First, data entered in the field is instantly synchronized with the server and available to other corporate users. Second, users do not need to install the software on their own computer, so upgrades and modifications can easily be made to the program and then quickly implemented. The computing

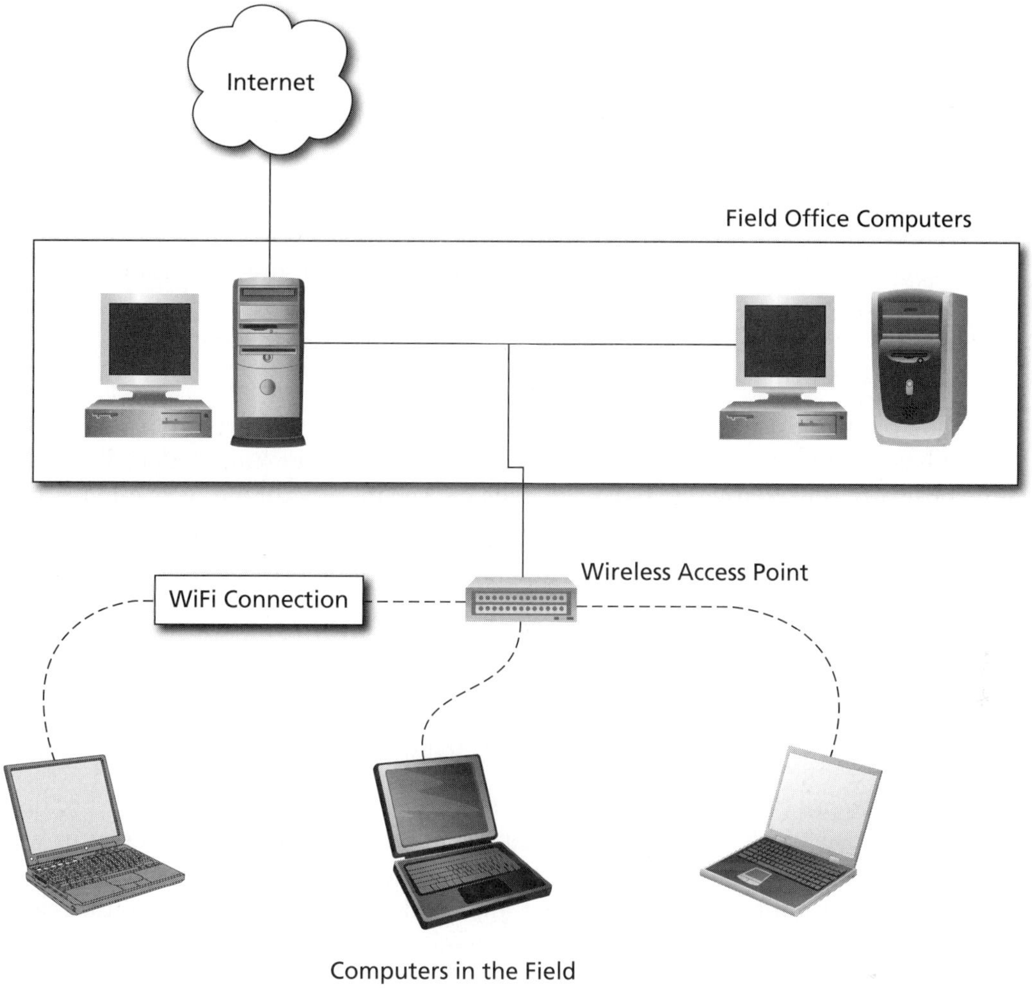

Figure 13.9 A typical Wi-Fi network. From WILLIAMS. *Information Technology for Construction Managers, Architects, and Engineers* 1E. © 2007 Delmar Learning, a part of Cengage Learning Inc. Reproduced by permission.

requirements of the mobile computer are reduced, reducing the size and cost of the mobile computer.

Web-based applications have already gained wide popularity through the use of programs such as Google Docs (docs.google.com), which has a word processor and spreadsheet or the Zoho family of online programs (www.zoho.com), which include a word processor, slide presentation software, note-talking software, and an online planner. Many Web-based programs are free to use.

Web-based applications for the construction industry range from complicated programs for managing construction projects and handling a huge number of documents to smaller-scale applications that a small contractor can implement to track project status and capture knowledge about a firm's best practices. In this section, two different types of applications will be discussed. First, weblogs will be described, followed by a discussion of a weblog that was used on a construction project. Then, a more complicated type of Internet-based application, the Web portal, will be described.

Weblogs

A **weblog** is composed of brief, frequently updated posts that are arranged chronologically (Bausch et al., 2002). No knowledge of computer programming or HTML is required to post information to a weblog. No software is required to be placed on a user's computer. Only Web access and a Web browser are required to access a weblog and add information (Stone, 2002).

Their ease of use makes weblogs an ideal application for the construction industry, where many people may not have extensive computer knowledge. Typically, weblog posts are automatically archived. This can be an advantage over e-mails because the message is archived in a central repository that everyone can access.

The weblog also can provide a means of fostering collaboration. There are various modes of using a weblog. One way is for a single person or entity to provide all the posts. Most weblog systems provide a feature for commenting. This allows weblog users to comment on posts that are made, and they are added to the weblog. An alternative approach that can allow for enhanced collaboration is to set up the weblog to have multiple posters. This can allow a dialog to develop between weblog users.

Weblogs can potentially find many uses in the construction industry because of their ease of use and flexibility. There has been considerable recent interest in weblogs for their capability to instantly transmit news and opinions. Some journalists publishing weblogs of political opinion have found wide audiences. In some industries, weblogs have been used as a collaborative tool where employees share and comment on ideas.

Weblog Functions and Capabilities

The primary feature of a weblog is the ability to post text onto a Web page. Typically, weblog posts are arranged chronologically and are given a time and a date stamp. In other words, the weblog provides a way to easily create Web content and send it to the WWW without requiring significant technical knowledge other than the knowledge of using a Web browser.

Weblogs can also be used to foster collaboration. Commenting is a feature that many weblog systems provide. Someone viewing the weblog can attach comments to the post, which are viewable by all weblog users. This feature allows weblogs to be used for collaboration and as a forum where ideas can be exchanged. In a weblog, posts are not discarded. Instead, an archive of older posts is created, which is accessible to the user. This is desirable for construction applications where the weblog can serve as a record of project incidents and milestones.

Weblogs typically have additional features. Primary among these capabilities is the ability to attach files to weblog posts so that they can be shared with other weblog users.

Another useful feature in the construction context is the ability to post photographs to a weblog. A natural use of this feature in construction is to provide and distribute progress photographs. Additionally, photographs of construction problems can be posted to a weblog to be rapidly sent to project team members.

A primary concern in the construction industry is to provide security for sensitive project data and the proprietary knowledge of the construction firm. The original weblogs that were developed did not have any security features. The weblogs were accessible by anyone who was able to locate the URL for the Web page. Weblogs are now available that provide security. Weblogs can be accessed in two ways. The first is to belong to a weblog service that provides hosting for a company's weblogs. The second way is to obtain weblog software and host the weblog on the firm's own computers. Even for a small company, it would be possible to host the weblogs using a personal computer as the server.

Both methods have advantages and drawbacks. Many construction companies have serious concerns about the security of their data. In working with a construction company with significant IT capabilities, it was found that they preferred to host their own weblog. Another concern with hosted services is that a stable hosting provider be selected. Some companies are small and may go out of business, with subsequent loss of project data. However, there are several well-known weblog service providers that do provide a safe haven for project data.

Of the weblog services available, many offer their services free of cost. However, the free services offer only basic features. Typically, these free services do not provide any security features for the Web site that is created. Anyone surfing the Web can view the weblog Web page. This lack of security typically rules out the use of free weblogs by construction companies for project management applications. However, weblog services that are provided for a fee typically provide password protection for the weblog Web site.

The original weblog service is Blogger. Anyone can establish a weblog free of cost at the Blogger site (www.blogger.com). Blogger has several powerful features, but it is limited in its construction applications because it does not offer passwords. There are many services available for a monthly fee, which provide adequate security and additional functions. Web links for some popular hosting services are shown in Table 13.1. There are many available and this list is by no means exhaustive.

Table 13.1 Weblog services and software

NAME	SERVICE	WEB SITE
Blogger	Weblog service	www.blogger.com
LiveJournal	Weblog service	www.livejournal.com
TypePad	Weblog service	www.typepad.com
MoveableType	Weblog software	www.moveabletype.org
Greymatter	Weblog software	www.noahgrey.com/greysoft

Weblog Example

To illustrate the uses of weblogs in construction, we present a description of a system of weblogs that was developed for managing the actual construction of a large high school project. The project's construction manager maintained these weblogs.

Several weblogs were used to assist in managing the construction project. These included a weblog of project progress photographs, a weblog of project meeting minutes, a calendar of upcoming meetings, a weblog of project financial data, and a weblog detailing outstanding project tasks. Figure 13.10 provides a table of contents page with links to the other weblogs.

To add information to the weblog, an editing window is displayed to the user. The user can type a text message in the window or paste a photo file to the window. No knowledge of any commands is required to post the information to the weblog page. Access to the SilkBlogs editor is Web based, so a content editor does not require any special software to make posts to the Web site. Text information can be cut or pasted from existing documents.

Figure 13.11 shows an example of a page from the task list weblog. It illustrates a typical weblog page with a chronological listing of posts. The task list weblog contains a list of outstanding project items. Posts are listed in chronological order and are automatically given a date and a time stamp. Each individual post can be expanded to show more detail. The weblog also contains a category area. A weblog post may be

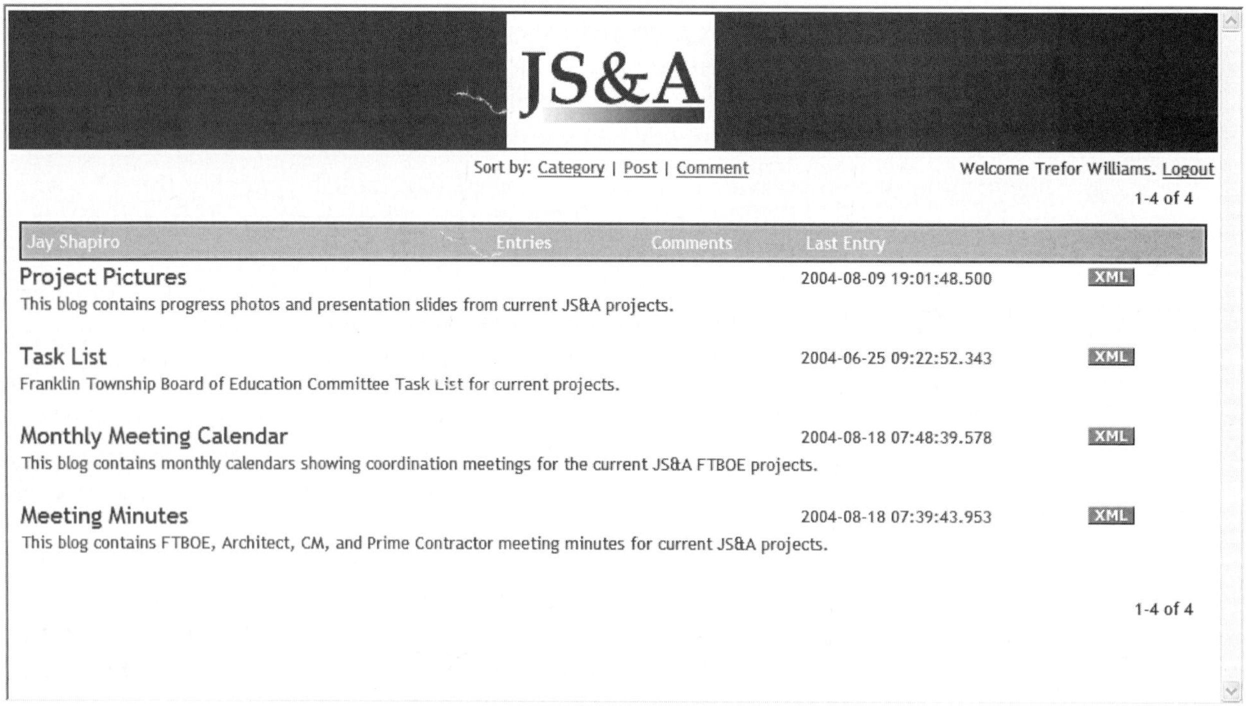

Figure 13.10 Weblog table of contents. *Courtesy of Jay Shapiro & Associates Inc.*

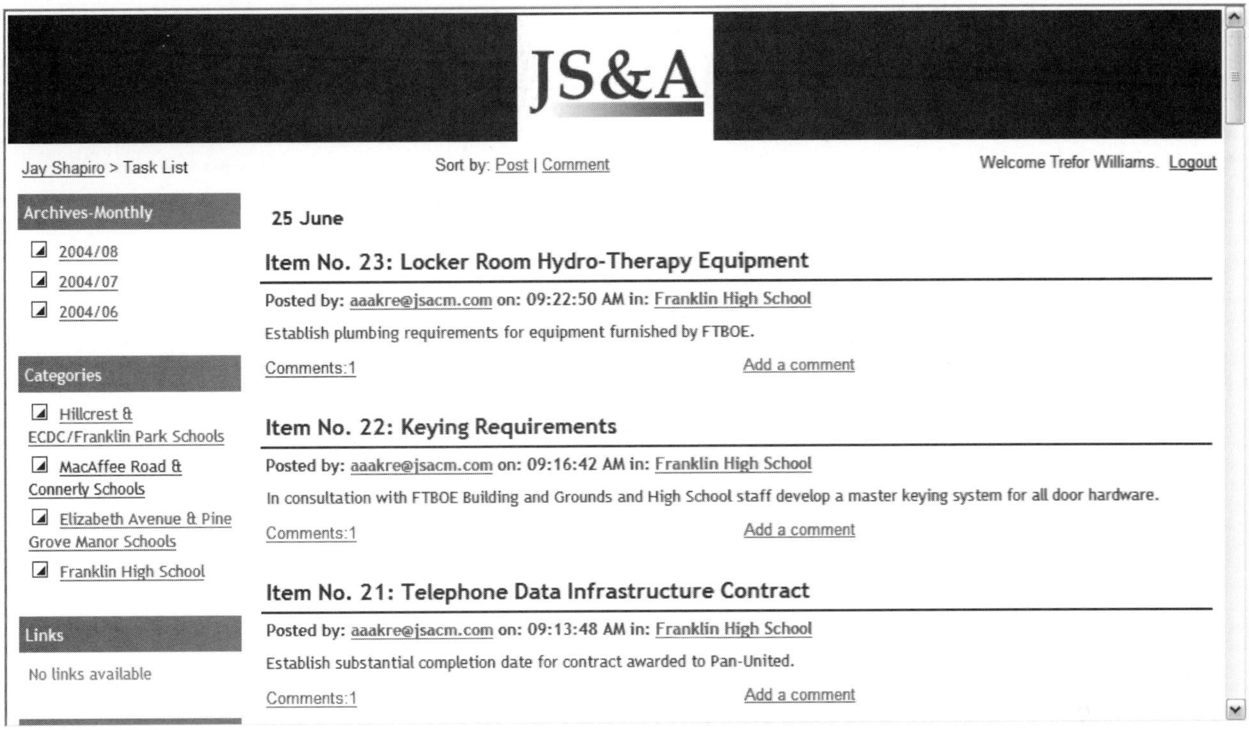

Figure 13.11 Weblog posts. *Courtesy of Jay Shapiro & Associates Inc.*

assigned to any category created by the user. In this case, the categories represent different projects being conducted for the client.

Figure 13.11 also illustrates how the commenting feature can be used for collaboration. Using the commenting feature of the weblog, project staffs commented on the status of the tasks and exchange information. Figure 13.12 shows a weblog post expanded to show the original post and a comment about the status of the task from another project engineer.

Several parties both within and outside the construction manager's organization used the weblogs. The average number of hits per day to the weblogs was 6.39. There were approximately 10 users. External organizations with access to the weblogs included the architect, subcontractors, and owner. It was found that the weblogs provided an alternative to e-mailing each participant project documents and messages. Using the weblog, users established two repositories to share documents with project participants. A weblog that contained the monthly invoices and a weblog that contained the minutes of project meetings were included in the weblog system. Figure 13.13 shows the weblog for project meetings where each individual post contains a link to the project meeting for a particular date. Figure 13.14 shows an individual post from the listing of monthly invoices. A link to the invoice PDF file is provided to the user who can download the document.

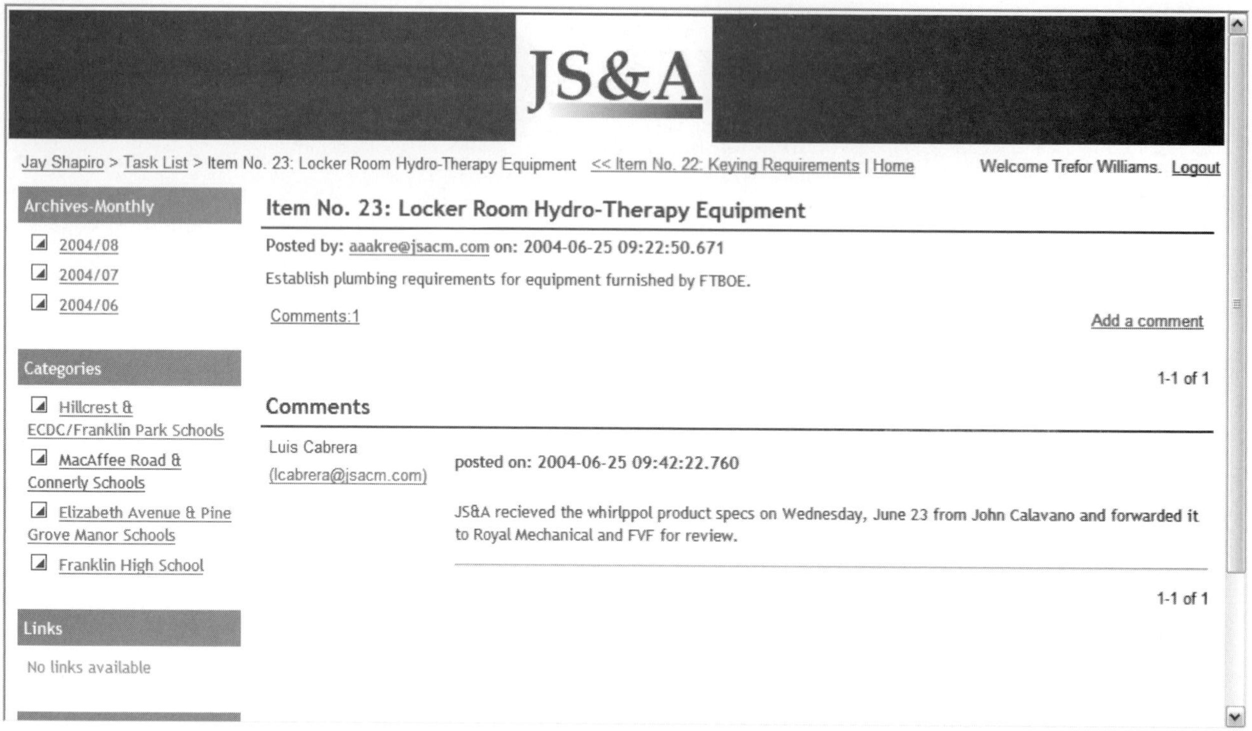

Figure 13.12 Web post with a comment. *Courtesy of Jay Shapiro & Associates Inc.*

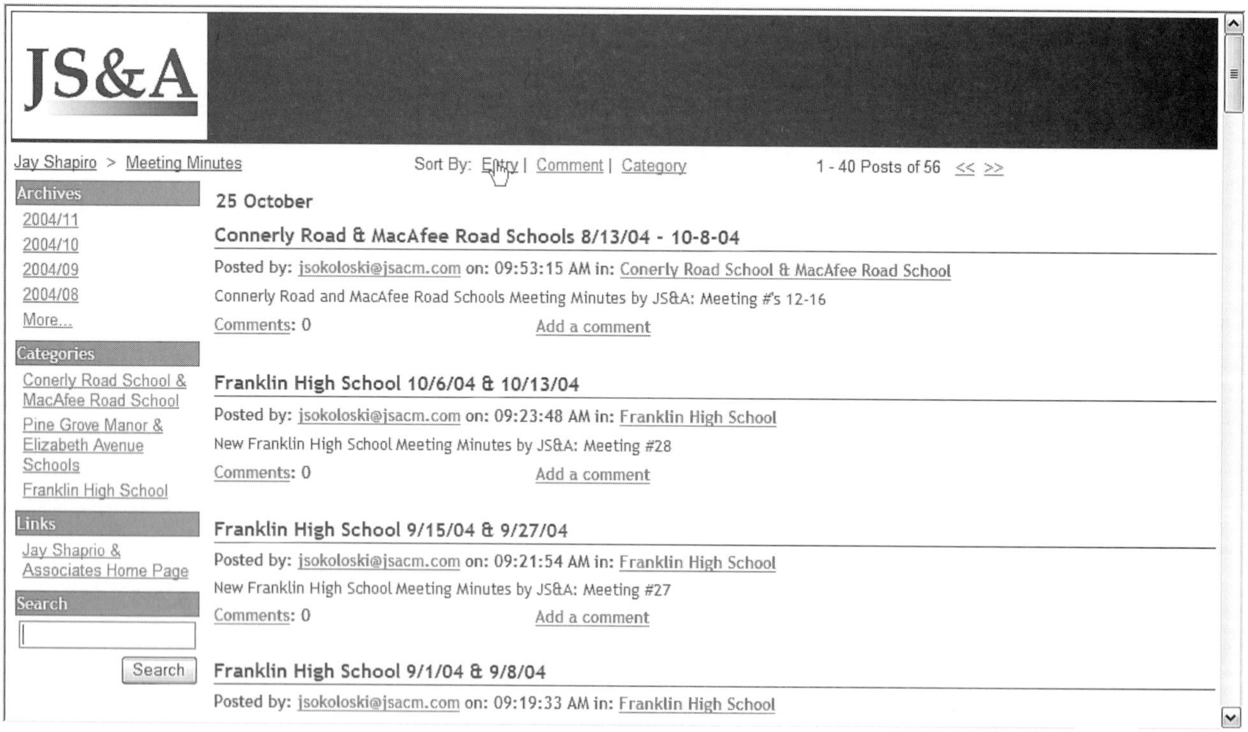

Figure 13.13 Weblog used to archive project meeting minutes. *Courtesy of Jay Shapiro & Associates Inc.*

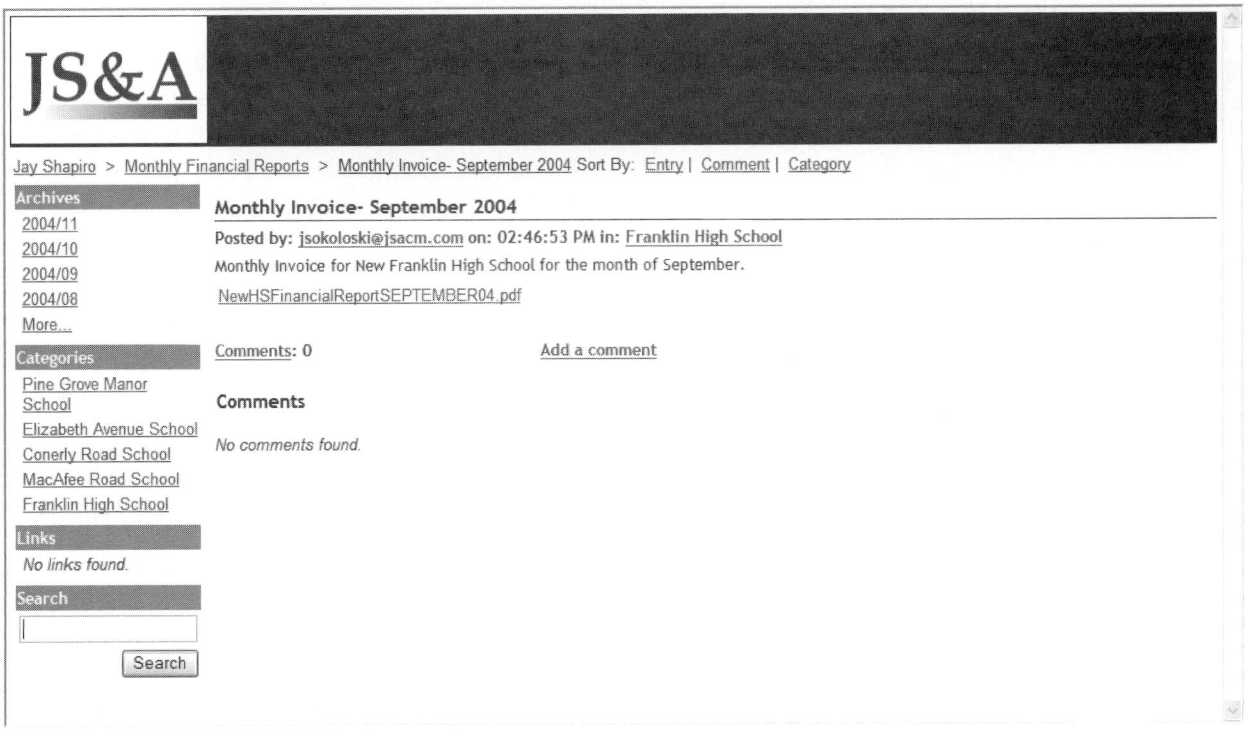

Figure 13.14 Link to a project document within a web post. *Courtesy of Jay Shapiro & Associates Inc.*

The photographic weblog provides a way for all project participants to observe project progress and to see pictures of project problem areas. Before implementation of the weblog, the construction manager had been preparing Microsoft PowerPoint presentations containing photographs of project progress for the owner. The weblog allowed the owner to access the progress information directly, resulting in a time saving for the project manager. It was also found that subcontractors were downloading the photographs to document their own work.

Success of the Weblog Implementation The weblogs developed provided a useful, time-saving tool for the high school project. Several elements were present that contributed to the success of this implementation. These include:

1. The attitude of the top management of the construction management firm was enthusiastic. The construction manager's top management viewed the implementation as a way of increasing the firm's prestige by demonstrating the firm's willingness to innovate to clients.
2. The construction manager provided resources to insure that the weblogs were implemented properly. An engineer was designated to implement and maintain the weblogs.
3. The implementation of the weblog was an incremental innovation. The construction manager was already using computers extensively to manage the project, and the Web-based weblogs were easily incorporated into the firm's day-to-day operations.

Web Portals

On complex projects, thousands of documents may be exchanged between project participants. Methods like a weblog are not sophisticated enough to handle the volume of paperwork that is produced on a complex project. A method to control, route, and modify project documents between the owner, designer, contractor, and subcontractors in a complex project is therefore required.

A Web portal (sometimes called a project extranet) is a web-based service that allows for collaboration and document exchange on construction projects. A Web portal provides a Web page for a construction project that allows users to access project documents and collaborate with each other through the exchange of messages.

The basic idea of a construction Web portal is to apply Web-based systems and appropriate construction project management techniques to facilitate online collaboration through better communication and workflow management. The Web portal thus provides efficient information exchange to assist in successfully delivering a project.

The primary reason for using a Web portal is to enhance communication between the project participants. In construction projects, poor communication is believed to impact projects significantly, by causing delays and inefficiencies. A Web portal allows key project information to be available to all participants and, if used properly, provides project participants with timely information about project status.

A possible benefit when employing a Web portal system is a decrease in the number of paper plans and documents that are required. Project participants can access project documents electronically through the Web portal and do not need to print documents.

A Central Point for Information Exchange

A construction project Web portal provides a central point where all project participants can interact. The many players in a construction contract, including the owner, designer, prime contractor, construction manager, and subcontractors, can all access the Web portal. In the absence of a Web portal, communications can be fragmented, with project participants communicating and exchanging documents by telephone, e-mail, mail, and fax. The Web portal provides a repository of all project documents.

A major benefit of Web portals is the enhanced integration of the design and the construction process. A Web portal for a project can be initialized during the design phase of the project, and all design information is available through it. During construction, required design changes and questions about the design can be effectively tracked through the Web portal. Communication between the designer and the contractor is enhanced and is faster than traditional, paper-based methods.

Types of Documents Exchanged

Different types of documents are exchanged on a typical construction project. On a large project, thousands of documents are generated. Web portals allow documents such as CAD files containing project plans, change orders, meeting minutes, requests for information, inspection reports, and shop drawings to be exchanged.

A major feature of Web portal systems is to provide document-tracking capabilities. Typically, Web portals can track who has viewed and modified documents. This can be particularly important on a complex project, where it may be necessary to track hundreds of change orders and requests for information simultaneously. The status of changes and questions can be quickly ascertained so that the project schedule is not impacted.

Web Portal Service Providers

Web portal software can be purchased or leased and installed on a construction company's own server. However, many contractors access Web portals as an online service provided by the Web portal company. In this way, contractors can access the sophisticated Web portal software over the Web, without needing to invest in significant computer infrastructure. It should also be noted that large contractors might need to subscribe to several Web portals simultaneously because contract specifications may require them to use a particular Web portal system on a project. Because of the many perceived benefits of using Web portals on complex projects, owners now often require the use of a particular Web portal system on a project.

There are many Web portal service providers. These include Expedition, Prolog, Constructware, and Buzzsaw. Companies that are well known in the construction industry provide some of the Web services. The authors of Primavera, the Primavera Project Planner scheduling software, provide a Web portal service. AutoDesk, the developers of AutoCAD, provides the Buzzsaw service and the Constructware service. Increasingly, Web portal service providers are providing links to other software. For example, the Prolog portal provides integration with Microsoft Word, Microsoft Project, Primavera Project Planner, and Bentley ProjectWise (Meridian Systems, 2005).

Construction Applications of Mobile and Wireless Computing

In this section, we will look at a sample of the applications that are available for use in the field. We will discuss applications for data collection, extensions to well-known software packages for CAD, and scheduling; we will also discuss possible knowledge management applications for mobile devices.

Data Collection Applications

Handheld computers have been successfully used in construction for field data collection. Some programs have been written for the activities of a construction contractor, whereas other field office software are available for use by government agencies, engineers, or architects supervising the construction of a project. Software aimed at construction contractors for use in field offices can organize project paperwork, including daily logs, inspection reports, time cards and cost control data. The software for project managers and owner organizations focus more on inspector reports and collection of pay item data. A major component of field office management programs is the capability to allow data collection in the field using a PDA or laptop computer, using forms that are easy to fill out in the field. Handheld computers are taken to the workplace to collect data and communicate to the field office computer wirelessly or are brought back to the field office and the data is downloaded to the office computer by a cradle and cable.

Quality Management Using a Blackberry

A company called ATSG (Advanced Technologies Support Group) has developed a data collection system for residential constructors to monitor construction quality. The system employs the Blackberry wireless handheld device, or a tablet PC, and the Construction Quality Manager software. The Blackberry device is popular because of its small size, ability to handle e-mails, and its full keyboard that makes data entry easier than on some other types of handheld devices. This software is for use by construction supervisors while walking through a house under construction. The construction supervisor records defects using the Blackberry device, and the recorded information is instantly sent to the central Construction Quality Manager database. Figure 13.15 shows a data input screen on a Blackberry device.

Figure 13.15 Entering punch list details on a Blackberry device. From WILLIAMS. *Information Technology for Construction Managers, Architects, and Engineers* 1E. © 2007 Delmar Learning, a part of Cengage Learning Inc. Reproduced by permission.

Subcontractors are automatically notified of defects and the required corrective actions via e-mail or fax. The Construction Quality Manager software is either installed on a company's network or hosted by ATSG. Grayson Homes, a homebuilder in the Baltimore/Washington area, uses the Construction Quality Manager software on its projects for quality assurance. The software provides a database and management reporting system that provides Grayson Homes with a comprehensive view of its business. The company also reports that field supervisors find the built-in capabilities of the Blackberry (e-mail, calendar, and address book) to be extremely useful when used along with the quality assurance software (Blackberry 2005).

OnSite Enterprise

AutoDesk sells a program called OnSite Enterprise that provides digital plans and maps from a central server to Pocket PC devices. A program called OnSite View allows a PDA device to operate both as a client when working with the OnSite Enterprise server and as a stand-alone device. When using the PDA as a stand-alone device, OnSite View plans can be loaded onto the PDA by syncing with a personal computer. This powerful application allows computerized plans and GIS files to be viewed at the point of work. These programs allow AutoCAD drawing format (DWG) and drawing interchange format files (DXF) format files to be viewed and marked up using a Pocket PC device. In addition, the OnSite software supports various GIS file formats.

The program uses the pen-based interface of the Pocket PC to allow users to pan and zoom in real time any view of the drawing. If the drawing contains layers, specified views that are saved within the drawing can be selected. In addition, the OnSite View program allows users to annotate plans by making free-hand drawings, using predefined symbols, and adding text. The markup files that are created on the Pocket PC can be moved back to a desktop PC and opened over a DWG or DXF file. Programs like this can allow easy reference to construction plans in the field. The ability to take notes and modify the drawings can allow rapid generation of as-built drawings. It can also provide a way of annotating construction problems in the field and marking their spatial location within the project (AutoDesk).

Mobile Scheduling

Mobile devices can be used in several ways to assist in scheduling. First, they can provide a way for personnel in the field to update activity information. Second, they allow managers at remote locations to access information about the status of projects. A good example of **mobile scheduling** is the mobile application that links to Primavera. The Primavera software has a handheld version called Mobile Manager. It runs on both the Pocket PC and Palm devices. The software allows managers to both view and modify scheduling data in the field. Mobile Manager has several useful features. These include:

- The ability to transfer data from a handheld device to a computer either wirelessly or by connecting physically to a computer running Primavera
- The feature by which users can select key items from a schedule for download to the handheld device

- The capability to download data from multiple projects to the mobile device
- The facility of single-screen accessibility and updating of activities, where a user can modify the activity in the field on the basis of observed conditions

Use of mobile devices for scheduling has several benefits, some of which are listed below:

- Data no longer have to be entered twice. They go directly from the handheld device into the company's scheduling system.
- Revised schedule data are more accurate because input comes from on-site observers and is recorded in real time.
- Managers outside the firm's office can access schedule details. Managers no longer have to return to the field office to determine the schedule status of construction operations.

Summary

Computers and the application of information technology are becoming ubiquitous in the construction industry. BIM can provide a powerful tool to design and construct buildings of all sizes. The use of BIM will continue to expand and all modern construction managers need knowledge of BIM to succeed in today's environment. IT allows knowledge capture and document management by providing Web-based tools and Web portal systems. Finally, wireless networks are allowing computers and mobile devices to provide managers in the field with mobile forms and the most up-to-date schedule and cost information. With the continual performance improvements generated, we can only expect to see more IT usage and more sophisticated and powerful applications of IT in construction.

Key Terms

Building information modeling (BIM)
Client/server computing
IT (information technology)
Mobile scheduling
Weblog

Review Questions

1. What is the difference between a weblog and a Web portal?
2. BIM is only a compilation of 2D CAD documents. T or F.
3. What is data interoperability? Why is it important?
4. BIM treats building components as objects that can be modified. T or F.
5. What is a LAN?

6. A construction company sets up a Wi-Fi network on a construction site. What benefits are achieved by providing wireless connectivity? Discuss the mobile applications that a Wi-Fi makes available to this project.

Management Pro

Make a weblog pertaining to a construction issue or topic. Make posts and add links pertinent to the topic. Add pictures or videos if possible. The weblog should be educational for others who view the weblog. A way to make a free weblog is to use a service like Blogger. Share your blog with other class members to get their comments about your informational weblog. Discuss your experience building and using the weblog. Do you think weblogs have a place in the management of construction projects?

References

Advanced Technologies Support Group. Construction quality manager. http://www.atsgi.com/PDF/CQM.pdf (accessed November 17, 2005).

AutoDesk. AutoDesk onsite enterprise overview. http://usa.autodesk.com/adsk/servlet/index?siteID=123112&id=700957 (accessed November 27, 2005).

Bausch, P., M. Haughey, and M. Houlihan. 2002. *We Blog: Publishing Online with Weblogs*. Indianapolis: Wiley.

Cory, Clark. 2001. Utilization of 2D, 3D, or 4D CAD in construction communication documentation. In: *Fifth International Conference on Information Visualisation*. pp. 219–224.

Dodd, Annabel. 2005. *The Essential Guide to Telecommunications*. Upper Saddle River, NJ: Pearson Education.

Eastman, Chuck. 2007. What is BIM? Altanta, GA. Available from http://bim.arch.gatech.edu/?id=402 (accessed May 15, 2008). BIM Resources@Georgia Tech.

Finith, Jernigan. 2007. Big BIM Little BIM: The Practical Approach to Building Information Modeling. Salisbury, MD: 4Site Press.

Gallaher, M.P., Alan C. O'Connor, John L. Dettbarn, Jr., and Linda T. Gilday. 2004. Cost Analysis of inadequate interoperability in the U.S. capital facilities industry. Gaithersburg, MD: National Institute of Standards and Technology, U.S. Department of Commerce.

Gallelo, Dominic, Marcel Broekmaat, and Clay Freeman. Virtual construction benefits. Available from http://www.graphisoft.com/products/construction/white_papers/whitepaper2.html (accessed December 1, 2005).

Gehry Technologies. 2005. Gehry Technologies announces major new software update. Available from http://www.gehrytechnologies.com/company-press-07-15-2005.html (accessed December 10, 2005).

Heesom, David, and Lamine Mahdjoubi. 2004. Trends of 4D CAD applications for construction planning. *Construction Management and Economics* 22: 171–182.

Lowe, Doug. 1999. *Client/Server Computing for Dummies*. Foster City, CA: IDG Books Worldwide.

Meridian Systems. 2005. Prolog manager overview. Available from http://www.mps.com/products/prolog/PM/index.asp (accessed August 29, 2005).

Sandsmark, Fred. 2004. What you need to know about wireless networking. *iQ Magazine* 5: 66–71.

Sheppard, Laurel. 2004. Virtual building for construction projects. *Computer Graphics and Applications* 24: 6–12.

Stone, B. 2002. *Blogging: Genius Strategies for Instant Web Content*. Indianapolis, IN: New Riders.

U.S. General Services Administration. 2007. GSA building information modeling guide series 01, version 0.60. Washington, D.C.: U.S. General Services Administration.

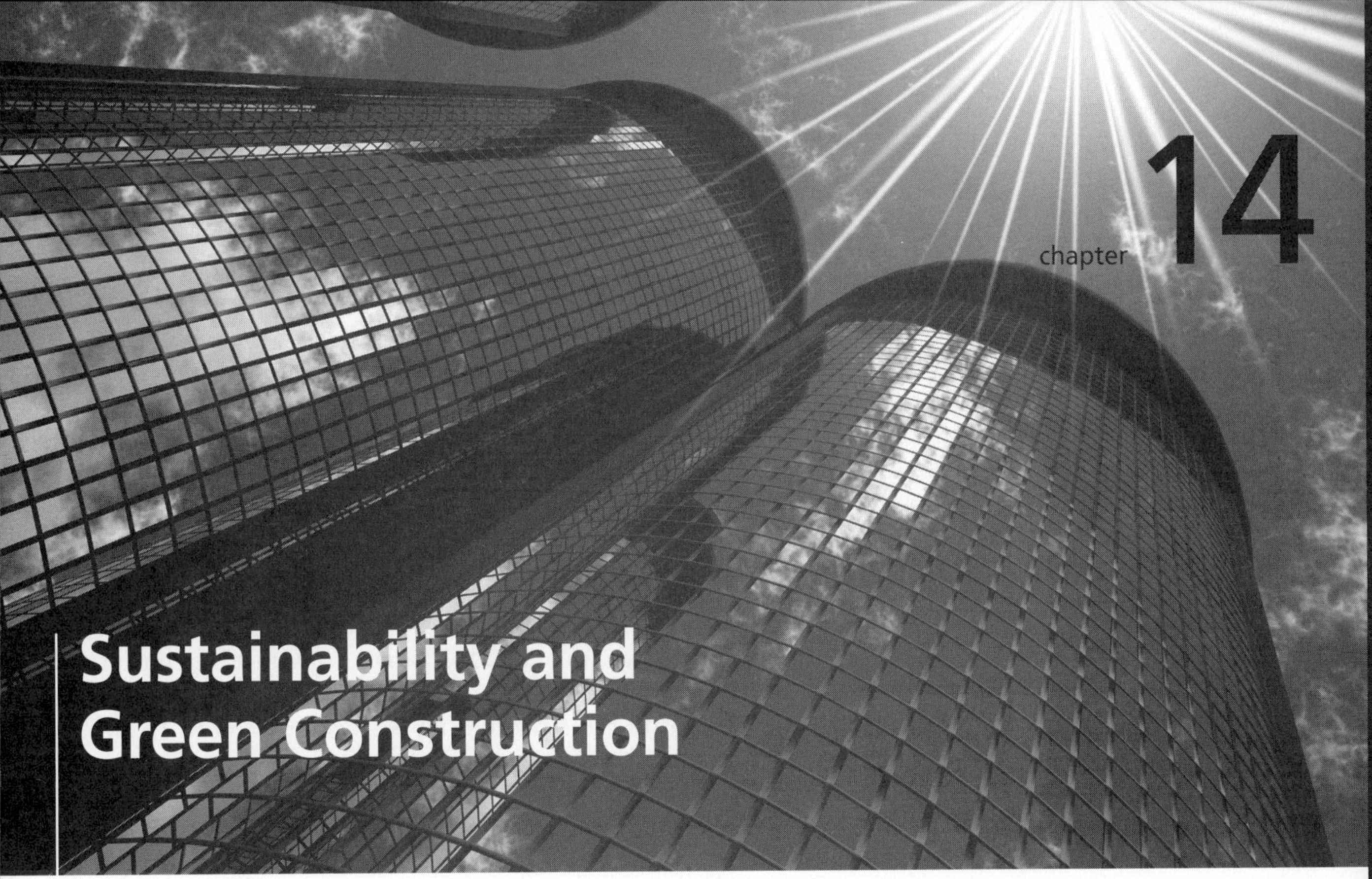

chapter 14

Sustainability and Green Construction

Chapter Outline

Introduction

The earth's resources are finite. In particular, petroleum and mineral resources are heavily used during construction and as components of many different types of facilities. There has been increasing recognition that it is necessary to seek methods of building that are sustainable to ensure that our standard of living can be maintained for future generations. In addition to a need to reduce reliance on nonrenewable natural resources, there are concerns about the environment. In particular, global warming from air pollution is a serious threat to the earth's ecosystems. Therefore, there are also concerns that we seek materials and methods that minimize the impacts of burning fossil fuels and the emission of harmful chemicals into the environment.

What Is Sustainability?

Sustainability is the concept of meeting the needs of the present without compromising the ability of future generations to meet their needs. For the construction contractor, sustainability means that new types of materials and techniques will be used to minimize the impact of construction on the surrounding ecosystems and to reduce consumption of materials that future generations will find difficult to replace. The construction contractor must understand that these new techniques will have profound effects on how construction is carried out and how much the construction costs.

Related to sustainability is the idea that construction materials are renewable or nonrenewable. Wood as a material is a renewable resource because new trees can be grown to replace the wood used on a construction project. Steel is a nonrenewable resource because it requires large quantities of nonrenewable coal and iron ore to produce. Although scrap steel can be recycled, it still requires large amounts of nonrenewable energy to be put in a useable form.

Sustainability Issues

The reasons for the increased interest in sustainability can be explained by several interconnected phenomena. First, the world is experiencing rapid population growth. This growth has spurred the use of nonrenewable resources. The increased use of some of these resources, then, has significant negative environmental impacts, such as the pollution produced by burning fossil fuels.

The world's population is continuing to grow, and countries such as China and India are rapidly industrializing, increasing the demand for natural resources. The world's population in 2010 is estimated to be 6,832,877,668. By 2050, the total population is projected to grow to 9,392,797,012. The world's most important natural resource is oil. We require petroleum products to fuel our vehicles and heat our homes; yet the burning of fossil fuels is the largest source of air pollution. The top oil user is the United States (17 million barrels per day). The top gas user comprises nations that constituted the former Soviet Union (23,000 billion cubic feet per year). The total world consumption of crude oil in 1996 was 71.7 million barrels per day. (There are 42 U.S. gallons in a barrel, or 159 liters.) OPEC estimates that the total world oil consumption could reach around 100 million barrels per day by the year 2020 (Offshore-environment.com, 2008). Today, oil supplies about 40% of the world's energy and 96% of its transportation energy. Since the shift from coal to oil, the world has consumed over 875 billion barrels. Another 1,000 billion barrels of proved and probable reserves remain to be recovered. From now to 2020, world oil consumption will rise by about 60%.

Transportation will be the fastest growing oil-consuming sector. By 2025, the number of cars will increase to well over 1.25 billion from approximately 700 million today. Global consumption of gasoline could double. The two countries with the highest rate of growth in oil use are China and India, whose combined populations account for a third of humanity. In the next two decades, China's oil consumption is expected to grow at a rate of 7.5% per year and India's 5.5% (compare with a 1% growth for the industrialized countries) (Institute for Analysis of Global Security, 2008). Clearly, these statistics show the need to embrace sustainability if resources are to be available for future generations.

These increasing demands for energy and the pressure of population growth generate different problems that must be addressed by the construction industry. Ten major impact categories of resource depletion and pollution have been identified by the National Institute of Standards and Technology (National Association of Home Builders, 2006). They are:

- Acid rain
- Ecological toxicity
- Eutrophication
- Global warming
- Human toxicity
- Indoor air quality
- Ozone depletion
- Resource depletion
- Smog
- Solid waste

Sustainable construction and building seeks to minimize a facility's contribution to these global problems over the facility's entire life cycle including construction.

Green Construction

Green construction is an organized effort to design and build buildings using a process and materials that promote environmental sustainability. Several approaches to green construction are possible. They include methods of resource conservation while building the facility, selection of materials that use fewer nonrenewable resources to produce, and the use of materials that can be recycled and reused on future projects.

Green Certification

Certification provides a framework for the implementation of green buildings and more sustainable projects. Typically, a project is certified as conforming to green design and construction standards by a third-party organization. A project will be certified if it meets the design and construction benchmarks of the certifying third-party organization. Green building rating systems are point based and require a project to earn a certain number of points before it can be certified as a green facility (Glavanich, 2008). The certification frameworks typically require a user to meet benchmarks involving a broad range of sustainability and environmental considerations.

The LEED System

LEED (Leadership in Energy and Environmental Design) is probably the best known certification program in the United States. LEED is a third-party certification program and a nationally accepted benchmark for the design, construction, and operation of high performance green buildings. LEED has been developed by the U.S. Green Building Council (USGBC), an industry organization with membership from all sections of the construction industry (Glavanich, 2008).

The LEED process is intended to impact the entire life cycle of a facility. The LEED process must be initiated for a project during the design phase, where green materials and techniques are selected that will impact how the project will be constructed. The process continues through construction, where efficient, nonpolluting practices are sought and continues through the operation of the constructed facility.

There are several LEED rating systems. They include rating systems for:

- New construction and major renovations of commercial buildings
- Core and shell
- Commercial interiors
- Existing buildings, including upgrades, operations, and maintenance

LEED promotes a whole-building approach to sustainability by recognizing performance in five key areas of human and environmental health: sustainable site development, water savings, energy efficiency, materials selection, and indoor environmental quality (U.S. Green Building Council, 2008). For a new commercial building, the LEED standard allows for the accumulation of up to 69 points for a project in these five key areas. A project is certified if it can accumulate between 26 and 32 points. There are silver, gold, and platinum levels of certification for projects that accumulate higher point totals.

Several benefits have been found in the construction of LEED-compliant buildings besides the reduction in waste and the creation of a more efficient building. Studies have found that LEED-certified buildings have a lower life cycle cost. It has been argued that energy efficiency, worker productivity, and other LEED features provide lifetime savings that are about 10 times greater than the initial cost premium of constructing the LEED certified facility (Soderland et al., 2008).

LEED Standards and Construction

Many LEED standards can directly affect the conduct of construction operations. These standards affect the methods of construction that are used as well as the materials that are used to construct the facility. In a sustainable building, a contractor will encounter new materials and new ways of handling materials to reduce negative impacts on the external environment and the environmental quality of the building's environment. One example from the LEED standard is Sustainable Sites Prerequisite 1 (U.S. Green Building Council, 2005). This is a standard to minimize soil erosion and dust generation from the construction site. The standard, which provides 1 point toward certification, requires an erosion and sediment control plan for all projects. The plan describes the measures required to:

- Prevent loss of soil during construction
- Prevent sedimentation in storm sewers or receiving streams
- Prevent polluting the air with soil and particulate matter

Strategies typically employed in an erosion control plan include temporary and permanent seeding and the use of basins and silt fences to capture eroding soil. A construction contractor must be aware of the techniques to be used and the cost impacts that more stringent specifications covering erosion control will have.

The LEED materials and resources MR 2.1 standard requires the recycling of 50% of the construction and demolition waste. Achieving this goal would provide two points in the LEED rating scheme. A construction waste management plan must be developed that identifies the materials to be diverted and if they will be sorted on or off site. Possible materials that can be recycled include cardboard, metal, brick, acoustical tile, concrete, plastic, clean wood, glass, gypsum wallboard, carpet, and insulation. Recyclers of these various materials must be identified. Specific areas on the construction site must be designated for collection of material to be recycled. Achieving this goal requires considerable planning by a construction contractor and represents a significant break from past practices. Costs will be greater and must be studied carefully by the contractor. The contractor will be required to work with area recyclers to find what materials can be recycled and where they must be shipped to.

The LEED standard also provides points for improvements to environmental quality. LEED EQ Credit 4.2: Low Emitting Materials: Paints and Coatings is an indoor environmental quality credit a project can receive if low-emitting paints and coatings are used to improve the indoor air quality. The purpose is to promote the use of paints and architectural coatings that do not emit large quantities of volatile gases into the building environment.

The designer will incorporate requirements of this type in the project specifications, and the contractor is responsible for changing construction methods to install the new types of materials. A standard of this type shows how some standards for sustainability will require contractors to handle and use new materials that have lower environmental impacts while other standards are concerned with practices such as recycling.

Example

The Bronx Library Center in New York City is an example of the benefits possible from LEED implementation. The project has a silver LEED certification. Ninety percent of demolition debris was recycled. There is a 20% energy saving in the building due to the use of natural light, sensors to lower or turn off lights in empty rooms, and roofing that reflects solar heat. Materials for the building were selected on the basis of environmental characteristics, with recycled content included in the structural steel, carpeting, and flooring. More than half of the building materials were manufactured within 500 miles of the project site to minimize the pollution caused when material is transported to the project sight.

Residential Green Home Building Guidelines

The LEED system focuses on commercial buildings. For residential construction, the National Association of Homebuilders (NAHB) has developed the Model Green Home Building Guidelines. These guidelines provide a framework for residential builders and architects to respond to clients seeking green construction and materials for their new home.

The NAHB guidelines are grouped into eight areas of guiding principles (National Association of Home Builders, 2006). The eight areas major areas are:

- **Lot design, preparation, and development.** This section includes methods for minimizing environmental impacts to the building site.
- **Resource efficiency.** This section comprises guidelines for the use of more efficient construction methods and materials.
- **Energy efficiency.** This section relates to guidelines to minimize energy use during construction and in the home after it is built.
- **Water efficiency guidelines.** The mean per capita indoor daily water use is 64 gallons. Homes designed using the water efficiency guidelines can reduce usage to less than 45 gallons.
- **Indoor environmental quality.** This includes guidelines that promote healthy indoor living conditions, including the reduction of emissions of unhealthy gases from building materials.
- **Maintenance.** This section involves guidelines to promote maintenance throughout the life of the home to maintain performance.
- **Global impact.** This section includes guidelines that address issues of global importance such as global warming.

- **Community site planning and land development.** Many home builders develop residential communities. These guidelines address ways to improve sustainability through the urban planning process.

Heavy Construction and Sustainability Certification

The use of certification to promote sustainability has focused on building construction. However, there are ongoing efforts to develop rating systems for heavy construction. One such effort is Green Roads, developed for use in Washington State Department of Transportation projects. The purpose of Green Roads is to construct more sustainable roadways. It contains guidelines for issues like pavement noise and light pollution that often arise in highway projects. It also contains guidelines for constructing cooler pavements to help reduce urban heat island effects.

Construction Materials and Sustainability

Sustainability has caused two major changes in construction. First, many materials that were not extensively recycled or reused in the past are now being reused in large quantities. Second, a large number of innovative materials and techniques are being developed that are more sustainable and use less energy than older materials.

Building materials that can be readily recycled or reused include:

- **Concrete.** Concrete can be recycled as compacted highway base-and-fill material.
- **Lumber.** There is a market for timber salvaged from demolished buildings. Lumber can be recycled as mulch or to make engineered wood products.
- **Asphalt roofing.** Asphalt roofing can be recycled to make hot mix asphalt, cold patch, fuel, or new roofing materials.

Other types of building products that are not recyclable can be reused if they are in a suitable condition (e.g., furnaces, plumbing fixtures, electrical fixtures, and window frames).

Gypsum wall board is an interesting example of the innovations that have occurred to promote sustainability. First, wall board made from recycled materials is now readily available. Second, new materials are available that allow drywall panels to be connected together with removable connections rather than taping the joints between panels. This allows the drywall to be easily moved and reused.

Another example of sustainable materials can be found in highway construction. Researchers are now developing warm mix asphalt (WMA) pavements to reduce the energy expended and pollution emitted in the manufacture of hot mix asphalt (HMA). WMA is asphalt that is produced at significantly lower temperatures than HMA. WMA has been found to reduce the emissions generated by an asphalt plant by up to 30–40%, for both sulfur dioxide and carbon dioxide. Other pollutants are reduced as well.

Less fuel is used in producing the asphalt. When paving, WMA has been found to have several benefits, including the ability to pave at lower temperatures, compared with HMA, and the ability to be transported longer distances (D'Angelo et al., 2008). This is an example of the innovation that is now underway to reduce the environmental impact of material production and improve sustainability of materials by using fewer nonrenewable resources.

Summary

This chapter has shown that green construction and sustainability are growing in importance. Contractors must realize that green construction will require new materials and methods. This chapter has discussed how green construction is currently implemented through rating schemes that include guidelines for increasing sustainability and improving environmental quality. Projects that accumulate enough points can be certified as green buildings. To implement a sustainable guidelines project, designers will write specifications requiring green practices and using new types of materials and methods. Contractors must understand how practices like recycling and the use of low environmental impact materials will affect project costs and required construction procedures. The increasing need for more sustainability will foster the development of an increasing number of innovative materials.

Key Terms

Certification
Green construction
LEED (Leadership in Energy and Environmental Design)
Sustainability

Review Questions

1. What is sustainability? Why is it now an important topic in the construction industry?
2. What is green certification?
3. What is warm mix asphalt? Why has it been developed?
4. List three building materials and describe how they can be recycled or reused.
5. What green certification program would be best suited to the construction of a commercial building?

Management Pro

Select a building material like concrete or drywall. Study the available sustainable forms of the material and alternative materials. How do they differ? Are they easily obtainable? How much more does the sustainable material cost? How will construction techniques change?

References

D'Angelo, John, E. Harm, J. Bartoszek, J. Cowsert, T. Harman, M. Jamshidi, W. Jones, D. Newcomb, B. Prowell, R. Sines, and B. Yeaton. 2008. Warm mix asphalt: European practice. Vol. FHWA-PL-08-007. Washington, D.C.: Federal Highway Administration.

Glavanich, Thomas E. 2008. *Contractor's Guide to Green Building Construction.* Hoboken, NJ: John Wiley.

Institute for Analysis of Global Security. 2008. The future of oil. Available from http://www.iags.org/futureofoil.html (accessed May 25, 2008).

National Association of Home Builders. 2006. *Model Green Home Building Guidelines.* Washington, D.C.: National Association of Home Builders.

Offshore-environment.com. 2008. Interesting facts about oil and gas and ocean environment. Available from http://www.offshore-environment.com/facts.html (accessed May 22, 2008).

Soderland, Martin, Stephen T. Muench, Kim A. Willoughby, Jeffrey S. Uhlmeyer, Jim Weston. 2008. Green roads: A sustainability rating system for roadways. Paper presented at the 87th Annual Meeting of the Transportation Research Board, Washington, D.C.

U.S. Green Building Council. 2005. LEED for new construction and major renovations. Washington, D.C.: U.S. Green Building Council.

U.S. Green Building Council. 2008. What is LEED? Available from http://www.usgbc.org/DisplayPage.aspx?CMSPageID=222 (accessed May 23, 2008).

Glossary

activity anything that consumes time

addenda design changes that are often made during the bidding period

agency CM a construction management contract in which a construction manager provides advisory services to the owner from the design stage of the project through completion, but is not contractually bound to the contractors performing the project

arbitration the hearing and determination of a dispute by an impartial referee agreed to by the owner and the construction contractor

automation the technique of making a device, machine, procedure, or process that is self-acting, self-moving, or fully autonomous

balance point the number of trucks that will balance the productivity of the loader; calculated to achieve a balance between the productivities of two equipment types

bar chart one of the basic scheduling techniques, a graphical depiction of the amount of time required and the sequence of construction activities; also called a Gantt chart

baseline the formal documenting of a product at some level of design; for a construction project, the project baseline describes the requirements and functions of the facility that is to be constructed

best value procurement a method of selecting a design-builder by combining consideration of cost and technical factors that best suits an owner's requirement; the winning proposer is not necessarily the lowest bidder or the bidder with highest technical expertise

bid bond a bond an owner will normally require contractors to submit with the bid, to provide bid security to the owner; this is done to protect the owner from financial loss if the winning low bidder cannot construct the project

bid estimates a cost estimate that includes the contractor's profit and overhead

bid opening the process in which the bids of all competing contractors are formally opened and the low bidder declared; traditionally, bids are submitted to a bid box in a location designated by the owner

bid package a collection of all the documents that contractors must consider when deciding to bid on a project

brownfield projects typically, projects built at a location where previously built structures or buildings existed

building information modeling (BIM) a process that uses virtual models, typically 3D or 4D models, whereby information and data can be linked to objects in the virtual model

build-operate-transfer (BOT) a PPP method in which a concession is granted to a consortium to construct a large infrastructure project for a period, usually between 10 and 30 years, at the end of which all operating rights and maintenance responsibilities revert to the government

cash flow the pattern of a company's income and expenditures and the resulting availability of cash

cash flow problem a problem generated when payments due to a construction contractor generally lag behind the contractor's expenditures on a project

certification a process that provides a framework for the implementation of green buildings and more sustainable projects

change management the process of reviewing and approving or denying changes to the project scope, schedule, and budget

change orders formal modifications of the original design, project duration, or cost

changed condition a condition encountered by a contractor that differs materially from what is indicated in the design documents or a condition encountered that is of an unusual nature that differs materially from conditions that are ordinarily encountered

claims unresolved conflicts between the owner and contractor

client/server computing a type of computer network with three components: a server computer, client computers, and a network that connects the server computer to the clients

CM at-risk (CMAR) a construction management contract in which a construction manager provides end-to-end project management services to the owner and is contractually liable for successful project completion

configuration management a technique to manage the change that occurs in complex projects to help keep projects on schedule and within budget

construction contractor a firm that constructs the project

construction manager a firm that provides professional services to an owner to assist in the planning, coordination, and construction of a facility

contract a document that defines the legal responsibilities of the parties to an agreement

cost index an index that may be used to project construction prices from the past to the present or future

critical activities activities that cannot be delayed without extending the project completion date

critical path method a network-based technique that determines a connected chain(s) of activities through a project that can have no scheduling leeway if the project is to finish at the earliest possible time

design-bid-build the traditional method of construction, where a project is completely designed, then put out to bid, and then constructed by the low bidder

design-build a contract in which a single entity is responsible for designing and building a construction project

design-build-finance-operate (DBFO) a PPP method similar to the BOT format, differing mainly in the manner in which the DBFO team is compensated

designer a firm that designs the project, typically producing plans and specifications for what is to be constructed

dispute resolution boards (DRBs) a method of resolving disagreements that occur during construction

early finish (EF) the earliest point at which an activity can be completed

early start (ES) the earliest time an activity may start

earned value (EV) total of the budgeted costs of a project's activities that have been completed to date

engineer's estimate a cost estimate that is a final check for the owner to determine if the project is financially feasible

equivalent grade a measure of total resistance; expressed in terms of percentage of grade

estimate an approximate calculation of the degree of work, that is, cost, of a construction project

experience modification rating (EMR) a statistical tool used by the insurance industry to determine insurance rates

fast-tracking the ability to compress the duration of a construction project by starting construction before the design is complete, typically implemented in the CMAR contract

finish-to-finish a relationship between two activities where both are designated to finish on the same day

float the commonly used term in the construction industry to refer to leeway

forward pass the first step in a CPM calculation, in which the early start and early finish of each activity are calculated

free float the amount of time for which an activity can be delayed without affecting the starting time of the following activities or the completion time of the project

general conditions conditions included in the bid package that define the contractual relationships between the owner and the contractor

globalization a situation where international borders are increasingly irrelevant and economic interdependencies between nations are heightened

grade resistance that component of the construction equipment weight that acts parallel to an inclined surface; increases on upgrades and decreases on downgrades

green construction an organized effort to design and build buildings using a process and materials that promote environmental sustainability

greenfield projects projects built on a new undisturbed site

gross domestic product (GDP) the total value of goods and services produced by the nation in a year

guaranteed maximum price (GMP) a contract in which the CMAR would be reimbursed for all direct project costs plus a fee up to the negotiated maximum price; if project costs were to exceed this amount, the CM would have to absorb the additional costs

"horse blanket" schedule a method typically used for the construction of highway or mass transit facilities where the work is spread out over a broad linear area

information technology (IT) a term that encompasses all forms of technology (computers, software, and telecommunications) used to create, store, exchange, and use information in its various forms

infrastructure the arrangement of a nation's civil and transportation networks and facilities

interoperability a set of specifications that allows for data exchange between software programs of different companies

lags used where the finish-to-start relationship is maintained but the following activity is delayed for some number of time periods specified by the scheduler

late finish (LF) the latest time an activity can finish without delaying the project finish

late start (LS) the latest time an activity may start without delaying the project finish

leadership in energy and environmental design (LEED) a well-known green certification program in the United States

leeway scheduling leeway, a term attributed to activities that are not on the critical path, meaning that their start times can be adjusted without affecting the completion date of the project

line of credit a loan from a bank provided to construction contractors to make up for the difference between their expenses and their revenues

liquidated damages damages or penalties for late completion of a project, usually indicated in the general conditions

litigation a contest authorized by law, in a court of justice, for the purpose of enforcing a right

mediation a negotiation to resolve differences that is conducted by some impartial party

megaprojects typically, huge infrastructure projects that require billions of dollars and several years to complete

mobile scheduling use of handheld mobile devices to provide field personnel and managers a way to update and access information from remote locations

mobilization often used by owner organizations as a means of reducing the overdraft loan to the contractor, because typically the owner's borrowing cost is lower than the contractor's borrowing cost

notice to proceed a document sent by the owner to the contractor that indicates that the project site is available and work may commence

Occupational Safety and Health Administration (OSHA) the federal agency responsible for workplace safety in the United States

order of magnitude a method to estimate cost, called a "ballpark" estimate, typically made at the initiation of a project before any plans have been drawn

overhead costs costs incurred by the contractor but not directly chargeable to any items physically installed on the project

owner a person, company, or agency for which the project is being constructed

partnering a voluntary organized process by which the parties to a construction contract perform as a team to achieve mutually beneficial goals; is a nonbinding process, not involving any contract document changes

planning the act of formulating a program for a course of action

plans Graphical representations of what is to be built

preliminary estimates cost estimates made during the design stage of the project to inform the owner about the expected project cost

prequalify the process by which some government agencies make contractors eligible to bid on the basis of the contractors' financial, managerial, and performance record; only contractors that the government agency prequalifies are allowed to bid on a project

productivity a measure of an output of construction work tasks

progress payments periodic payments made to the contractor during construction

project an endeavor undertaken in an organization to create a new product, and completed within the constraints of time and resources

proposal form the document used by the contractor to specify his or her bid prices

public–private partnerships (PPPs) an arrangement in which government agencies lacking financial resources partner with the private sector in the whole life cycle of public works projects, including the financing of projects

quantity takeoff a measurement and calculation from the plans of the quantities of work that need to be performed on a project

request for proposals (RFP) a document prepared by the owner in which the owner defines the requirements of the project, and must provide a clear understanding of the scope of work and the required outcomes

retainage funds withheld from progress payments to the contractor until the end of the project as an incentive for project completion

retrofitting incorporation of newer systems or technologies to existing (older) buildings

right of way land over which a highway or transportation facility is constructed

rimpull the maximum amount of pull that can be generated at the wheel rims of a wheeled vehicle

robot typically, a moving machine that operates automatically after programming

rolling resistance the external force that resists the motion of wheeled vehicles

scheduling the preparation of a timetable for the completion of the various stages of a complex project

scope creep the cumulative effect of many seemingly small changes that accumulate over time and change the scope of the project into something larger and costlier than what was originally intended; configuration management is designed to eliminate scope creep

self-insure the capability of some large construction companies by which, instead of taking a policy from an insurance company, they pay premiums into an escrow account and administer the claims themselves

shadow tolls payments made by the host government to the contractor on the basis of traffic flows at predetermined points along the roadway

special conditions conditions that cover the unique requirements of a particular project that are not covered by the general conditions

specifications textual descriptions of general information and quality requirements of items shown in the plans

start-to-start a relationship between activities where both are designated to start on the same day

subcontractor a firm that the prime construction contractor hires to complete some portions of the work, usually in an extremely complex project requiring a lot of expertise

sustainability the concept of meeting the needs of the present without compromising the ability of future generations to meet their needs

time extension additional time given to a contractor to complete work if problems beyond the contractor's control have occurred that delayed the project

total float the amount of time an activity can be delayed without delaying project completion; some following activities may be delayed if an activity uses total float; it is a widely used term, because the scheduling software that is most commonly used in the industry provides total float as an integral part of their outputs

total resistance the sum of rolling resistance and grade resistance

unbalanced bid a method adopted by contractors to reduce the amount of the overdraft, in which unit prices are readjusted to earn more revenue in the early stages of the project

unit pricing a process that determines how much each unit will cost to produce, transport, and install in the correct position as required by the project

value engineering a systematic review of a project involving use of creative thinking by a team of experts to improve performance, quality, and life cycle costs

weblog a Web page comprising brief, frequently updated posts that are arranged chronologically

workers compensation insurance paid by the contractor, to compensate workers who have been injured in the workplace

Index

Note: Page numbers followed by *f* indicate figure.